www.EffortlessMath.com

... So Much More Online!

✓ FREE Math lessons

✓ More Math learning books!

✓ Mathematics Worksheets

✓ Online Math Tutors

Need a PDF version of this book?

Visit www.EffortlessMath.com

Or send email to: info@EffortlessMath.com

AFOQT Math Prep 2019

Prep 2019

A Comprehensive Review and Ultimate Guide to the AFOQT Math Test

By

Reza Nazari & Ava Ross

All inquiries should be addressed to:

info@effortlessMath.com

www.EffortlessMath.com

ISBN-13: 978-1-970036-02-2

ISBN-10: 1-970036-02-8

Published by: Effortless Math Education

www.EffortlessMath.com

Description

AFOQT Math Prep 2019 provides students with the confidence and math skills they need to succeed on the AFOQT Math, building a solid foundation of basic Math topics with abundant exercises for each topic. It is designed to address the needs of AFOQT test takers who must have a working knowledge of basic Math.

This comprehensive book with over 2,500 sample questions and 10 complete AFOQT Arithmetic Reasoning and Mathematics knowledge tests is all you need to fully prepare for the AFOQT Math. It will help you learn everything you need to ace the math section of the AFOQT.

Effortless Math unique study program provides you with an in-depth focus on the math portion of the test, helping you master the math skills that students find the most troublesome.

This book contains most common sample questions that are most likely to appear in the mathematics section of the AFOQT.

Inside the pages of this comprehensive AFOQT Math book, students can learn basic math operations in a structured manner with a complete study program to help them understand essential math skills. It also has many exciting features, including:

- Dynamic design and easy-to-follow activities
- A fun, interactive and concrete learning process
- Targeted, skill-building practices
- Fun exercises that build confidence
- Math topics are grouped by category, so you can focus on the topics you struggle on
- All solutions for the exercises are included, so you will always find the answers
- 10 Complete AFOQT Math Practice Tests that reflect the format and question types on AFOQT

AFOQT Math Prep 2019 is an incredibly useful tool for those who want to review all topics being covered on the AFOQT test. It efficiently and effectively reinforces learning outcomes through engaging questions and repeated practice, helping you to quickly master basic Math skills.

About the Author

Reza Nazari is the author of more than 100 Math learning books including:
– **Math and Critical Thinking Challenges:** For the Middle and High School Student
– **GED Math in 30 Days**
– **ASVAB Math Workbook 2018 - 2019**
– **Effortless Math Education Workbooks**
– **and many more Mathematics books ...**

Reza is also an experienced Math instructor and a test–prep expert who has been tutoring students since 2008. Reza is the founder of Effortless Math Education, a tutoring company that has helped many students raise their standardized test scores—and attend the colleges of their dreams. Reza provides an individualized custom learning plan and the personalized attention that makes a difference in how students view math.

You can contact Reza via email at:
reza@EffortlessMath.com

Find Reza's professional profile at:
goo.gl/zoC9rJ

Contents

Chapter 1: Whole Numbers

Topics that you'll learn in this chapter:

- ✓ Place Value
- ✓ Rounding
- ✓ Whole Number Addition and Subtraction
- ✓ Whole Number Multiplication and Division
- ✓ Rounding and Estimates
- ✓ Comparing Numbers

"If people do not believe that mathematics is simple, it is only because they do not realize how complicated life is." — John von Neumann

Place Value

| Helpful Hints | The value of the place, or position, of a digit in a number.
For the number 3,684.26 | Example:

In 456, the 5 is in "tens" position. |

Decimal Place Value Chart

Millions	Hundred thousands	Ten thousands	Thousands	Hundreds	Tens	Ones	Decimal point	Tenths	Hundredths	Thousandths	Ten-thousandths	Hundred thousandths	Millionths
			3	6	8	4	.	2	6				

✎ *Write each number in expanded form.*

1) Thirty–five   30 + 5

2) Sixty–seven   ___ + ___

3) Forty–two   ___ + ___

4) Eighty–nine   ___ + ___

5) Ninety–one   ___ + ___

✎ *Circle the correct one.*

6)   The 2 in 72 is in the   ones place   tens place   hundreds place

7)   The 6 in 65 is in the   ones place   tens place   hundreds place

8)   The 2 in 342 is in the   ones place   tens place   hundreds place

9)   The 5 in 450 is in the   ones place   tens place   hundreds place

10)   The 3 in 321 is in the   ones place   tens place   hundreds place

Rounding

Helpful *Hints*	– Rounding is putting a number up or down to the nearest whole number or the nearest hundred, etc.	**Example:** 64 rounded to the nearest ten is 60, because 64 is closer to 60 than to 70.

✍️ *Round each number to the underlined place value.*

1) <u>9</u>72

2) 2,<u>9</u>95

3) 3<u>6</u>4

4) <u>8</u>1

5) <u>5</u>5

6) 33<u>4</u>

7) 1,<u>2</u>03

8) 9.<u>5</u>7

9) 7.<u>4</u>84

10) 9.1<u>4</u>

11) <u>3</u>9

12) <u>9</u>,123

13) 3,4<u>5</u>2

14) <u>5</u>69

15) 1,<u>2</u>30

16) <u>9</u>8

17) <u>9</u>3

18) <u>3</u>7

19) 4<u>9</u>3

20) 2,<u>9</u>23

21) <u>9</u>,845

22) 5<u>5</u>5

23) <u>9</u>39

24) <u>6</u>9

Whole Number Addition and Subtraction

Helpful Hints	1– Line up the numbers.	Example:
	2– Start with the unit place. (ones place)	231 + 120 = 351
	3– Regroup if necessary.	292 – 90 = 202
	4– Add or subtract the tens place.	
	5– Continue with other digits.	

✏ *Solve.*

1) A school had 891 students last year. If all last year students and 338 new students have registered for this year, how many students will there be in total?

2) Alice has just started her first job after graduating from college. Her yearly income is $33,000 per year. Alice's father income is $56,000 per year and her mother's income is $49,000. What is yearly income of Alice and her parent altogether?

3) Tom had $895 dollars in his saving account. He gave $235 dollars to his sister, Lisa. How much money does he have left?

4) Emily has 830 marbles, Daniel has 970 marbles, and Ethan has 230 marbles less than Daniel. How many marbles do they have in all?

✏ *Find the missing number.*

5) 890 – .... = 300

6) 1000 – .... = 200

7) .... – 4000 = 92000

8) 60000 – 51000 = ...

9) 3400 – .... = 3200

10) 33000 – 5000 = ...

Whole Number Multiplication and Division

Helpful	Multiplication:	Example:
	– Learn the times tables first!	$200 \times 90 = 18,000$
Hints	– For multiplication, line up the numbers you are multiplying.	
	– Start with the ones place.	$18,000 \div 90 = 200$
	– Continue with other digits	
	– A typical division problem:	
	Dividend ÷ Divisor = Quotient	
	Division:	
	– In division, we want to find how many times a number (divisor) is contained in another number (dividend).	
	– The result in a division problem is the quotient.	

✎ *Multiply and divided.*

1) $340 \div 8 =$

2) $1800 \div 20 =$

3) $50000 \div 10 =$

4) $966 \div 30 =$

5) $201 \times 20 =$

6) $400 \times 50 =$

7) $400 \times 90 =$

8) $888 \times 90 =$

9) $80 \times 80 =$

10) $122 \times 12 =$

11) $609 \times 8 =$

12) $220 \times 12 =$

13) A group of 235 students has collected $8,565 for charity during last month. They decided to split the money evenly among 5 charities. How much will each charity receive?

14) Maria and her two brothers have 9 boxes of crayons. Each box contains 56 crayons. How many crayons do Maria and her two brothers have?

Rounding and Estimates

Helpful *Hints*	– Rounding and estimating are math strategies used for approximating a number. – To estimate means to make a rough guess or calculation. – To round means to simplify a known number by scaling it slightly up or down.	**Example:** $73 + 69 \approx 140$

✎*Estimate the sum by rounding each added to the nearest ten.*

1) 55 + 9

2) 25 + 12

3) 83 + 7

4) 32 + 37

5) 13 + 74

6) 34 + 11

7) 39 + 77

8) 25 + 4

9) 61 + 73

10) 64 + 59

11) 14 + 68

12) 82 + 12

13) 43 + 66

14) 45 + 65

15) 553 + 232

16) 418 + 846

17) 582 + 277

18) 2771 + 1651

19) 7436 + 3575

20) 1542 + 8738

21) 3843 + 6579

22) 4722 + 8186

23) 2419 + 7224

24) 6768 + 3169

Comparing Numbers

Helpful	Comparing:	Example:
	Equal to =	
Hints	Less than <	56 > 35
	Greater than >	
	Greater than or equal ≥	
	Less than or equal ≤	

✍ *Use > = <.*

1) 35	67	8) 90	56
2) 89	56	9) 94	98
3) 56	35	10) 48	23
4) 27	56	11) 24	54
5) 34	34	12) 89	89
6) 28	45	13) 50	30
7) 89	67	14) 20	20

✍ *Use less than, equal to or greater than.*

15) 23 _____ 34	22) 56 _____ 43
16) 89 _____ 98	23) 34 _____ 34
17) 45 _____ 25	24) 92 _____ 98
18) 34 _____ 32	25) 38 _____ 46
19) 91 _____ 91	26) 67 _____ 58
20) 57 _____ 55	27) 88 _____ 69
21) 85 _____ 78	28) 23 _____ 34

Answers of Worksheets – Chapter 1

Place Value

1) 30 + 5
2) 60 + 7
3) 40 + 2
4) 80 + 9
5) 90 + 1
6) ones place
7) tens place
8) ones place
9) tens place
10) hundreds place

Rounding

1) 1000
2) 3000
3) 360
4) 80
5) 60
6) 330
7) 1200
8) 9.6
9) 7.5
10) 9.1
11) 40
12) 9000
13) 3,450
14) 600
15) 1,200
16) 100
17) 90
18) 40
19) 490
20) 2,900
21) 10,000
22) 560
23) 900
24) 70

Whole Number Addition and Subtraction

1) 1229
2) 138000
3) 660
4) 2540
5) 590
6) 800
7) 96000
8) 9000
9) 200
10) 28000

Whole Number Multiplication and Division

1) 42.5
2) 90
3) 5000
4) 32.2
5) 4020
6) 20000
7) 36000
8) 79920
9) 6400
10) 1464
11) 4872
12) 2640
13) 1713
14) 504

Rounding and Estimates

1) 70	9) 130	17) 860
2) 40	10) 120	18) 4420
3) 90	11) 80	19) 11020
4) 70	12) 90	20) 10280
5) 37	13) 110	21) 10420
6) 40	14) 120	22) 12910
7) 120	15) 780	23) 9640
8) 30	16) 1270	24) 9940

Comparing Numbers

1) 35 < 67	15) 23 less than 34
2) 89 > 56	16) 89 less than 98
3) 56 > 35	17) 45 greater than 25
4) 27< 56	18) 34 greater than 32
5) 34 = 34	19) 91 equal to 91
6) 28 < 45	20) 57 greater than 55
7) 89 > 67	21) 85 greater than 78
8) 90 > 56	22) 56 greater than 43
9) 94 < 98	23) 34 equal to 34
10) 48 > 23	24) 92 less than 98
11) 24 < 54	25) 38 less than 46
12) 89 = 89	26) 67 greater than 58
13) 50 > 30	27) 88 greater than 69
14) 20 = 20	28) 23 less than 34

Chapter 2: Fractions and Decimals

Topics that you'll learn in this chapter:

- ✓ Simplifying Fractions
- ✓ Adding and Subtracting Fractions
- ✓ Multiplying and Dividing Fractions
- ✓ Adding Mixed Numbers
- ✓ Subtract Mixed Numbers
- ✓ Multiplying Mixed Numbers
- ✓ Dividing Mixed Numbers
- ✓ Comparing Decimals
- ✓ Rounding Decimals

- ✓ Adding and Subtracting Decimals
- ✓ Multiplying and Dividing Decimals
- ✓ Converting Between Fractions, Decimals and Mixed Numbers
- ✓ Factoring Numbers
- ✓ Greatest Common Factor
- ✓ Least Common Multiple
- ✓ Divisibility Rules

"A Man is like a fraction whose numerator is what he is and whose denominator is what he thinks of himself. The larger the denominator, the smaller the fraction." —Tolstoy

Simplifying Fractions

Helpful		Example:
Helpful	– Evenly divide both the top and bottom of the fraction by 2, 3, 5, 7, … etc. – Continue until you can't go any further.	
Hints		$\frac{4}{12} = \frac{2}{6} = \frac{1}{3}$

✍ *Simplify the fractions.*

1) $\frac{22}{36}$

2) $\frac{8}{10}$

3) $\frac{12}{18}$

4) $\frac{6}{8}$

5) $\frac{13}{39}$

6) $\frac{5}{20}$

7) $\frac{16}{36}$

8) $\frac{18}{36}$

9) $\frac{20}{50}$

10) $\frac{6}{54}$

11) $\frac{45}{81}$

12) $\frac{21}{28}$

13) $\frac{35}{56}$

14) $\frac{52}{64}$

15) $\frac{13}{65}$

16) $\frac{44}{77}$

17) $\frac{21}{42}$

18) $\frac{15}{36}$

19) $\frac{9}{24}$

20) $\frac{20}{80}$

21) $\frac{25}{45}$

Adding and Subtracting Fractions

Helpful

Hints

– For "like" fractions (fractions with the same denominator), add or subtract the numerators and write the answer over the common denominator.

– Find equivalent fractions with the same denominator before you can add or subtract fractions with different denominators.

– Adding and Subtracting with the same denominator:

$$\frac{a}{b} + \frac{c}{b} = \frac{a+c}{b}$$

$$\frac{a}{b} - \frac{c}{b} = \frac{a-c}{b}$$

– Adding and Subtracting fractions with different denominators:

$$\frac{a}{b} + \frac{c}{d} = \frac{ad+cb}{bd}$$

$$\frac{a}{b} - \frac{c}{d} = \frac{ad-cb}{bd}$$

✍ **Add fractions.**

1) $\frac{2}{3} + \frac{1}{2}$

2) $\frac{3}{5} + \frac{1}{3}$

3) $\frac{5}{6} + \frac{1}{2}$

4) $\frac{7}{4} + \frac{5}{9}$

5) $\frac{2}{5} + \frac{1}{5}$

6) $\frac{3}{7} + \frac{1}{2}$

7) $\frac{3}{4} + \frac{2}{5}$

8) $\frac{2}{3} + \frac{1}{5}$

9) $\frac{16}{25} + \frac{3}{5}$

✍ **Subtract fractions.**

10) $\frac{4}{5} - \frac{2}{5}$

11) $\frac{3}{5} - \frac{2}{7}$

12) $\frac{1}{2} - \frac{1}{3}$

13) $\frac{8}{9} - \frac{3}{5}$

14) $\frac{3}{7} - \frac{3}{14}$

15) $\frac{4}{15} - \frac{1}{10}$

16) $\frac{3}{4} - \frac{13}{18}$

17) $\frac{5}{8} - \frac{2}{5}$

18) $\frac{1}{2} - \frac{1}{9}$

Multiplying and Dividing Fractions

Helpful *Hints*	– **Multiplying fractions:** multiply the top numbers and multiply the bottom numbers. – **Dividing fractions:** Keep, Change, Flip Keep first fraction, change division sign to multiplication, and flip the numerator and denominator of the second fraction. Then, solve!	**Example:** $\dfrac{a}{b} \times \dfrac{c}{d} = \dfrac{a \times c}{b \times d}$ $\dfrac{a}{b} \div \dfrac{c}{d} = \dfrac{a}{b} \times \dfrac{d}{c} = \dfrac{ad}{bc}$

✏️ **Multiplying fractions. Then simplify.**

1) $\dfrac{1}{5} \times \dfrac{2}{3}$

2) $\dfrac{3}{4} \times \dfrac{2}{3}$

3) $\dfrac{2}{5} \times \dfrac{3}{7}$

4) $\dfrac{3}{8} \times \dfrac{1}{3}$

5) $\dfrac{3}{5} \times \dfrac{2}{5}$

6) $\dfrac{7}{9} \times \dfrac{1}{3}$

7) $\dfrac{2}{3} \times \dfrac{3}{8}$

8) $\dfrac{1}{4} \times \dfrac{1}{3}$

9) $\dfrac{5}{7} \times \dfrac{7}{12}$

✏️ **Dividing fractions.**

10) $\dfrac{2}{9} \div \dfrac{1}{4}$

11) $\dfrac{1}{2} \div \dfrac{1}{3}$

12) $\dfrac{6}{11} \div \dfrac{3}{4}$

13) $\dfrac{11}{14} \div \dfrac{1}{10}$

14) $\dfrac{3}{5} \div \dfrac{5}{9}$

15) $\dfrac{1}{2} \div \dfrac{1}{2}$

16) $\dfrac{3}{5} \div \dfrac{1}{5}$

17) $\dfrac{12}{21} \div \dfrac{3}{7}$

18) $\dfrac{5}{14} \div \dfrac{9}{10}$

Adding Mixed Numbers

Helpful Hints	Use the following steps for both adding and subtracting mixed numbers.	Example:
	− Find the Least Common Denominator (LCD)	
	− Find the equivalent fractions for each mixed number.	$1\frac{3}{4} + 2\frac{3}{8} = 4\frac{1}{8}$
	− Add fractions after finding common denominator.	
	− Write your answer in lowest terms.	

✎ *Add.*

1) $4\frac{1}{2} + 5\frac{1}{2}$

2) $2\frac{3}{8} + 3\frac{1}{8}$

3) $6\frac{1}{5} + 3\frac{2}{5}$

4) $1\frac{1}{3} + 2\frac{2}{3}$

5) $5\frac{1}{6} + 5\frac{1}{2}$

6) $3\frac{1}{3} + 1\frac{1}{3}$

7) $1\frac{10}{11} + 1\frac{1}{3}$

8) $2\frac{3}{6} + 1\frac{1}{2}$

9) $5\frac{3}{5} + 5\frac{1}{5}$

10) $7 + \frac{1}{5}$

11) $1\frac{5}{7} + \frac{1}{3}$

12) $2\frac{1}{4} + 1\frac{1}{2}$

Subtract Mixed Numbers

| Helpful | Use the following steps for both adding and subtracting mixed numbers. | Example: |

Helpful

Hints

Use the following steps for both adding and subtracting mixed numbers.

Find the Least Common Denominator (LCD)
- Find the equivalent fractions for each mixed number.
- Add or subtract fractions after finding common denominator.
- Write your answer in lowest terms.

Example:

$$5\frac{2}{3} - 3\frac{2}{7} = 2\frac{8}{21}$$

✎ *Subtract.*

1) $4\frac{1}{2} - 3\frac{1}{2}$

5) $6\frac{1}{6} - 5\frac{1}{2}$

9) $6\frac{3}{5} - 2\frac{1}{5}$

2) $3\frac{3}{8} - 3\frac{1}{8}$

6) $3\frac{1}{3} - 1\frac{1}{3}$

10) $7\frac{2}{5} - 1\frac{1}{5}$

3) $6\frac{3}{5} - 5\frac{1}{5}$

7) $2\frac{10}{11} - 1\frac{1}{3}$

11) $2\frac{5}{7} - 1\frac{1}{3}$

4) $2\frac{1}{3} - 1\frac{2}{3}$

8) $2\frac{1}{2} - 1\frac{1}{2}$

12) $2\frac{1}{4} - 1\frac{1}{2}$

Multiplying Mixed Numbers

Helpful	1- Convert the mixed numbers to improper fractions.	**Example:**
	2- Multiply fractions and simplify if necessary.	$2\frac{1}{3} \times 5\frac{3}{7} =$
Hints	$a\frac{c}{b} = a + \frac{c}{b} = \frac{ab+c}{b}$	$\frac{7}{3} \times \frac{38}{7} = \frac{38}{3} = 12\frac{2}{3}$

✍ **Find each product.**

1) $1\frac{2}{3} \times 1\frac{1}{4}$

2) $1\frac{3}{5} \times 1\frac{2}{3}$

3) $1\frac{2}{3} \times 3\frac{2}{7}$

4) $4\frac{1}{8} \times 1\frac{2}{5}$

5) $2\frac{2}{5} \times 3\frac{1}{5}$

6) $1\frac{1}{3} \times 1\frac{2}{3}$

7) $1\frac{5}{8} \times 2\frac{1}{2}$

8) $3\frac{2}{5} \times 2\frac{1}{5}$

9) $2\frac{2}{3} \times 4\frac{1}{4}$

10) $2\frac{3}{5} \times 1\frac{2}{4}$

11) $1\frac{1}{3} \times 1\frac{1}{4}$

12) $3\frac{2}{5} \times 1\frac{1}{5}$

Dividing Mixed Numbers

Helpful	1- Convert the mixed numbers to improper fractions.	**Example:**
Hints	2- Divide fractions and simplify if necessary.	$10\frac{1}{2} \div 5\frac{3}{5} =$

$$a\frac{c}{b} = a + \frac{c}{b} = \frac{ab+c}{b}$$

$$\frac{21}{2} \div \frac{28}{5} = \frac{21}{2} \times \frac{5}{28} = \frac{105}{56}$$

$$= 1\frac{7}{8}$$

✎ *Find each quotient.*

1) $2\frac{1}{5} \div 2\frac{1}{2}$

2) $2\frac{3}{5} \div 1\frac{1}{3}$

3) $3\frac{1}{6} \div 4\frac{2}{3}$

4) $1\frac{2}{3} \div 3\frac{1}{3}$

5) $4\frac{1}{8} \div 2\frac{2}{4}$

6) $3\frac{1}{2} \div 2\frac{3}{5}$

7) $3\frac{5}{9} \div 1\frac{2}{5}$

8) $2\frac{2}{7} \div 1\frac{1}{2}$

9) $3\frac{1}{5} \div 1\frac{1}{2}$

10) $4\frac{3}{5} \div 2\frac{1}{3}$

11) $6\frac{1}{6} \div 1\frac{2}{3}$

12) $2\frac{2}{3} \div 1\frac{1}{3}$

Comparing Decimals

Helpful	-	**Decimals:** is a fraction written in a special form. For example, instead of writing $\frac{1}{2}$ you can write 0.5.	**Example:**
Hints	-	**For comparing:** Equal to = Less than < Greater than > Greater than or equal ≥ Less than or equal ≤	2.67 > 0.267

✎*Write the correct comparison symbol (>, < or =).*

1) 1.25   2.3

2) 0.5   0.23

3) 3.2   3.2

4) 4.58   45.8

5) 2.75   0.275

6) 5.2   5

7) 3.1   0.31

8) 6.33   0.733

9) 8   0.8

10) 4.56   0.456

11) 1.12   1.14

12) 2.77   2.78

13) 6.08   6.11

14) 1.11   0.211

15) 2.6   2.55

16) 1.24   1.25

17) 5.52   0.552

18) 0.33   0.033

19) 14.4   14.4

20) 0.05   0.50

21) 0.59   0.7

22) 0.5   0.05

23) 0.90   0.9

24) 0.27   0.4

Rounding Decimals

Helpful Hints	We can round decimals to a certain accuracy or number of decimal places. This is used to make calculation easier to do and results easier to understand, when exact values are not too important.	**Example:** $6.37 = 6$

First, you'll need to remember your place values:

12.4567

1: tens	2: ones	4: tenths
5: hundredths	6: thousandths	7: tens thousandths

✎ *Round each decimal number to the nearest place indicated.*

1) 0.2̲3

2) 4.0̲4

3) 5.6̲23

4) 0.2̲66

5) 6̲.37

6) 0.8̲8

7) 8.2̲4

8) 7̲.0760

9) 1.62̲9

10) 6.3̲959

11) 1̲.9

12) 5̲.2167

13) 5.8̲63

14) 8.5̲4

15) 80̲.69

16) 65̲.85

17) 70.7̲8

18) 615̲.755

19) 16̲.4

20) 95̲.81

21) 2̲.408

22) 76̲.3

23) 116.5̲14

24) 8.0̲6

Adding and Subtracting Decimals

Helpful	1– Line up the numbers.	Example:
Hints	2– Add zeros to have same number of digits for both numbers.	16.18 $- \ 13.45$
	3– Add or Subtract using column addition or subtraction.	$\overline{2.73}$

✍ **Add and subtract decimals.**

1) $\begin{array}{r} 15.14 \\ - \ 12.18 \\ \hline \end{array}$

3) $\begin{array}{r} 82.56 \\ + \ 12.28 \\ \hline \end{array}$

5) $\begin{array}{r} 90.37 \\ + \ 56.97 \\ \hline \end{array}$

2) $\begin{array}{r} 65.72 \\ + \ 43.67 \\ \hline \end{array}$

4) $\begin{array}{r} 34.18 \\ - \ 23.45 \\ \hline \end{array}$

6) $\begin{array}{r} 45.78 \\ - \ 23.39 \\ \hline \end{array}$

✍ **Solve.**

7) _____ + 1.3 = 4.8

8) 4.2 + _____ = 11.6

9) 9.9 + _____ = 16

10) 6.9 + _____ = 16.4

11) _____ + 5.1 = 8.6

12) _____ + 7.9 = 15.2

Multiplying and Dividing Decimals

Helpful	**For Multiplication:**
Hints	– Set up and multiply the numbers as you do with whole numbers.
	– Count the total number of decimal places in both of the factors.
	– Place the decimal point in the product.
	For Division:
	– If the divisor is not a whole number, move decimal point to right to make it a whole number. Do the same for dividend.
	– Divide similar to whole numbers.

✍️ **Find each product.**

1) $\begin{array}{r} 4.5 \\ \times\ 1.6 \\ \hline \end{array}$

4) $\begin{array}{r} 8.9 \\ \times\ 9.7 \\ \hline \end{array}$

7) $\begin{array}{r} 5.7 \\ \times\ 7.8 \\ \hline \end{array}$

2) $\begin{array}{r} 7.7 \\ \times\ 9.9 \\ \hline \end{array}$

5) $\begin{array}{r} 15.1 \\ \times\ 12.6 \\ \hline \end{array}$

8) $\begin{array}{r} 98.20 \\ \times\ 100 \\ \hline \end{array}$

3) $\begin{array}{r} 2.6 \\ \times\ 1.5 \\ \hline \end{array}$

6) $\begin{array}{r} 6.9 \\ \times\ 3.3 \\ \hline \end{array}$

9) $\begin{array}{r} 23.99 \\ \times\ 1000 \\ \hline \end{array}$

✍️ **Find each quotient.**

10) $9.2 \div 3.6$

11) $27.6 \div 3.8$

12) $12.6 \div 4.7$

13) $6.5 \div 8.1$

14) $1.4 \div 10$

15) $3.6 \div 100$

16) $4.24 \div 10$

17) $14.6 \div 100$

18) $1.8 \div 1000$

Converting Between Fractions, Decimals and Mixed Numbers

Helpful	**Fraction to Decimal:**
	– Divide the top number by the bottom number.
Hints	**Decimal to Fraction:**
	– Write decimal over 1.
	– Multiply both top and bottom by 10 for every digit on the right side of the decimal point.
	– Simplify.

✍️ Convert fractions to decimals.

1) $\dfrac{9}{10}$

2) $\dfrac{56}{100}$

3) $\dfrac{3}{4}$

4) $\dfrac{2}{5}$

5) $\dfrac{3}{9}$

6) $\dfrac{40}{50}$

7) $\dfrac{12}{10}$

8) $\dfrac{8}{5}$

9) $\dfrac{69}{10}$

✍️ Convert decimal into fraction or mixed numbers.

10) 0.3

11) 4.5

12) 2.5

13) 2.3

14) 0.8

15) 0.25

16) 0.14

17) 0.2

18) 0.08

19) 0.45

20) 2.6

21) 5.2

Factoring Numbers

Helpful	-	Factoring numbers means to break the numbers into their prime factors.	**Example:**
Hints	-	First few prime numbers: 2, 3, 5, 7, 11, 13, 17, 19	$12 = 2 \times 2 \times 3$

✍ **List all positive factors of each number.**

1) 68

2) 56

3) 24

4) 40

5) 86

6) 78

7) 50

8) 98

9) 45

10) 26

11) 54

12) 28

13) 55

14) 85

15) 48

✍ **List the prime factorization for each number.**

16) 50

17) 25

18) 69

19) 21

20) 45

21) 68

22) 26

23) 86

24) 93

Greatest Common Factor

Helpful *Hints*	- List the prime factors of each number. - Multiply common prime factors.	**Example:** $200 = 2 \times 2 \times 2 \times 5 \times 5$ $60 = 2 \times 2 \times 3 \times 5$ $GCF\ (200, 60) = 2 \times 2 \times 5 = 20$

✎ *Find the GCF for each number pair.*

1) 20, 30

2) 4, 14

3) 5, 45

4) 68, 12

5) 5, 12

6) 15, 27

7) 3, 24

8) 34, 6

9) 4, 10

10) 5, 3

11) 6, 16

12) 30, 3

13) 24, 28

14) 70, 10

15) 45, 8

16) 90, 35

17) 78, 34

18) 55, 75

19) 60, 72

20) 100, 78

21) 30, 40

Least Common Multiple

Helpful	-	Find the GCF for the two numbers.	**Example:**
Hints	-	Divide that GCF into either number.	
	-	Take that answer and multiply it by the other number.	LCM (200, 60):
			GCF is 20
			$200 \div 20 = 10$
			$10 \times 60 = 600$

✎ *Find the LCM for each number pair.*

1) 4, 14

2) 5, 15

3) 16, 10

4) 4, 34

5) 8, 3

6) 12, 24

7) 9, 18

8) 5, 6

9) 8, 19

10) 9, 21

11) 19, 29

12) 7, 6

13) 25, 6

14) 4, 8

15) 30, 10, 50

16) 18, 36, 27

17) 12, 8, 18

18) 8, 18, 4

19) 26, 20, 30

20) 10, 4, 24

21) 15, 30, 45

Divisibility Rules

Helpful Hints	-	Divisibility means that a number can be divided by other numbers evenly.	**Example:** 24 is divisible by 6, because 24 ÷ 6 = 4

✍ *Use the divisibility rules to find the factors of each number.*

8	<u>2</u> 3 <u>4</u> 5 6 7 <u>8</u> 9 10
1) 16	2 3 4 5 6 7 8 9 10
2) 10	2 3 4 5 6 7 8 9 10
3) 15	2 3 4 5 6 7 8 9 10
4) 28	2 3 4 5 6 7 8 9 10
5) 36	2 3 4 5 6 7 8 9 10
6) 15	2 3 4 5 6 7 8 9 10
7) 27	2 3 4 5 6 7 8 9 10
8) 70	2 3 4 5 6 7 8 9 10
9) 57	2 3 4 5 6 7 8 9 10
10) 102	2 3 4 5 6 7 8 9 10
11) 144	2 3 4 5 6 7 8 9 10
12) 75	2 3 4 5 6 7 8 9 10

Answers of Worksheets – Chapter 2

Simplifying Fractions

1) $\dfrac{11}{18}$

2) $\dfrac{4}{5}$

3) $\dfrac{2}{3}$

4) $\dfrac{3}{4}$

5) $\dfrac{1}{3}$

6) $\dfrac{1}{4}$

7) $\dfrac{4}{9}$

8) $\dfrac{1}{2}$

9) $\dfrac{2}{5}$

10) $\dfrac{1}{9}$

11) $\dfrac{5}{9}$

12) $\dfrac{3}{4}$

13) $\dfrac{5}{8}$

14) $\dfrac{13}{16}$

15) $\dfrac{1}{5}$

16) $\dfrac{4}{7}$

17) $\dfrac{1}{2}$

18) $\dfrac{5}{12}$

19) $\dfrac{3}{8}$

20) $\dfrac{1}{4}$

21) $\dfrac{5}{9}$

Adding and Subtracting Fractions

1) $\dfrac{7}{6}$

2) $\dfrac{14}{15}$

3) $\dfrac{4}{3}$

4) $\dfrac{83}{36}$

5) $\dfrac{3}{5}$

6) $\dfrac{13}{14}$

7) $\dfrac{23}{20}$

8) $\dfrac{13}{15}$

9) $\dfrac{31}{25}$

10) $\dfrac{2}{5}$

11) $\dfrac{11}{35}$

12) $\dfrac{1}{6}$

13) $\dfrac{13}{45}$

14) $\dfrac{3}{14}$

15) $\dfrac{1}{6}$

16) $\dfrac{1}{36}$

17) $\dfrac{9}{40}$

18) $\dfrac{7}{18}$

Multiplying and Dividing Fractions

1) $\dfrac{2}{15}$

2) $\dfrac{1}{2}$

3) $\dfrac{6}{35}$

4) $\dfrac{1}{8}$

5) $\dfrac{6}{25}$

6) $\dfrac{7}{27}$

7) $\dfrac{1}{4}$

8) $\dfrac{1}{12}$

9) $\dfrac{5}{12}$

10) $\dfrac{8}{9}$

11) $\dfrac{3}{2}$

12) $\dfrac{8}{11}$

13) $\dfrac{55}{7}$

14) $\dfrac{27}{25}$

15) 1

16) 3

17) $\dfrac{4}{3}$

18) $\dfrac{25}{63}$

Adding Mixed Numbers

1) 10

2) $5\dfrac{1}{2}$

3) $9\dfrac{3}{5}$

4) 4

5) $10\dfrac{2}{3}$

6) $4\dfrac{2}{3}$

7) $3\dfrac{8}{33}$

8) 4

9) $10\dfrac{4}{5}$

10) $7\dfrac{1}{5}$

11) $2\dfrac{1}{21}$

12) $3\dfrac{3}{4}$

Subtract Mixed Numbers

1) 1

2) $\dfrac{1}{4}$

3) $1\dfrac{2}{5}$

4) $\dfrac{2}{3}$

5) $\dfrac{2}{3}$

6) 2

7) $1\dfrac{19}{33}$

8) 1

9) $4\dfrac{2}{5}$

10) $6\dfrac{1}{5}$

11) $1\dfrac{8}{21}$

12) $\dfrac{3}{4}$

Multiplying Mixed Numbers

1) $2\frac{1}{12}$

2) $2\frac{2}{3}$

3) $5\frac{10}{21}$

4) $5\frac{31}{40}$

5) $7\frac{17}{25}$

6) $2\frac{2}{9}$

7) $4\frac{1}{16}$

8) $7\frac{12}{25}$

9) $11\frac{1}{3}$

10) $3\frac{9}{10}$

11) $1\frac{2}{3}$

12) $4\frac{2}{25}$

Dividing Mixed Numbers

1) $\frac{22}{25}$

2) $1\frac{19}{20}$

3) $\frac{19}{28}$

4) $\frac{1}{2}$

5) $1\frac{13}{20}$

6) $1\frac{9}{26}$

7) $2\frac{34}{63}$

8) $1\frac{11}{21}$

9) $2\frac{2}{15}$

10) $1\frac{34}{35}$

11) $3\frac{7}{10}$

12) 2

Comparing Decimals

1) 1.25 < 2.3

2) 0.5 > 0.23

3) 3.2 = 3.2

4) 4.58 < 45.8

5) 2.75 > 0.275

6) 5.2 > 5

7) 3.1 > 0.31

8) 6.33 > 0.733

9) 8 > 0.8

10) 4.56 > 0.456

11) 1.12 < 1.14

12) 2.77 < 2.78

13) 6.08 < 6.11

14) 1.11 > 0.211

15) 2.6 > 2.55

16) 1.24 < 1.25

17) 5.52 > 0.552

18) 0.33 > 0.033

19) 14.4 = 14.4

20) 0.05 < 0.50

21) 0.59 < 0.7

22) 0.5 > 0.05

23) 0.90 = 0.9

24) 0.27 < 0.4

Rounding Decimals

1) 0.2
2) 4.0
3) 5.6
4) 0.3
5) 6
6) 0.9
7) 8.2
8) 7

9) 1.63
10) 6.4
11) 2
12) 5
13) 5.9
14) 8.5
15) 81
16) 66

17) 70.8
18) 616
19) 16
20) 96
21) 2
22) 76
23) 116.5
24) 8.1

Adding and Subtracting Decimals

1) 2.96
2) 109.39
3) 94.84
4) 10.73

5) 147.34
6) 22.39
7) 3.5
8) 7.4

9) 6.1
10) 9.5
11) 3.5
12) 7.3

Multiplying and Dividing Decimals

1) 7.2
2) 76.23
3) 3.9
4) 86.33
5) 190.26
6) 22.77

7) 44.46
8) 9820
9) 23990
10) 2.5555...
11) 7.2631...
12) 2.6808...

13) 0.8024...
14) 0.14
15) 0.036
16) 0.424
17) 0.146
18) 0.0018

Converting Between Fractions, Decimals and Mixed Numbers

1) 0.9
2) 0.56
3) 0.75
4) 0.4
5) 0.333...
6) 0.8

7) 1.2
8) 1.6
9) 6.9
10) $\frac{3}{10}$
11) $4\frac{1}{2}$

12) $2\frac{1}{2}$
13) $2\frac{3}{10}$
14) $\frac{4}{5}$
15) $\frac{1}{4}$

16) $\frac{7}{50}$ 18) $\frac{2}{25}$ 20) $2\frac{3}{5}$

17) $\frac{1}{5}$ 19) $\frac{9}{20}$ 21) $5\frac{1}{5}$

Factoring Numbers

1) 1, 2, 4, 17, 34, 68
2) 1, 2, 4, 7, 8, 14, 28, 56
3) 1, 2, 3, 4, 6, 8, 12, 24
4) 1, 2, 4, 5, 8, 10, 20, 40
5) 1, 2, 43, 86
6) 1, 2, 3, 6, 13, 26, 39, 78
7) 1, 2, 5, 10, 25, 50
8) 1, 2, 7, 14, 49, 98
9) 1, 3, 5, 9, 15, 45
10) 1, 2, 13, 26
11) 1, 2, 3, 6, 9, 18, 27, 54
12) 1, 2, 4, 7, 14, 28

13) 1, 5, 11, 55
14) 1, 5, 17, 85
15) 1, 2, 3, 4, 6, 8, 12, 16, 24, 48
16) 2 × 5 × 5
17) 5 × 5
18) 3 × 23
19) 3 × 7
20) 3 × 3 × 5
21) 2 × 2 × 17
22) 2 × 13
23) 2 × 43
24) 3 × 31

Greatest Common Factor

1) 10
2) 2
3) 5
4) 4
5) 1
6) 3
7) 3
8) 2
9) 2
10) 1
11) 2
12) 3
13) 4
14) 10
15) 1
16) 5
17) 2
18) 5
19) 12
20) 2
21) 10

Least Common Multiple

1) 28
2) 15
3) 80
4) 68
5) 24
6) 24
7) 18
8) 30
9) 152
10) 63
11) 551
12) 42
13) 150
14) 8
15) 150
16) 108
17) 72
18) 72
19) 780
20) 120
21) 90

Divisibility Rules

1) 16

<u>2</u> 3 <u>4</u> 5 6 7 <u>8</u> 9 10

2) 10

<u>2</u> 3 4 <u>5</u> 6 7 8 9 <u>10</u>

3) 15

2 <u>3</u> 4 <u>5</u> 6 7 8 9 10

4) 28

<u>2</u> 3 <u>4</u> 5 6 <u>7</u> 8 9 10

5) 36

<u>2</u> <u>3</u> <u>4</u> 5 <u>6</u> 7 8 <u>9</u> 10

6) 18

<u>2</u> <u>3</u> 4 5 6 <u>6</u> 7 8 <u>9</u> 10

7) 27

2 <u>3</u> 4 5 6 7 8 <u>9</u> 10

8) 70

<u>2</u> 3 4 <u>5</u> 6 <u>7</u> 8 9 <u>10</u>

9) 57

2 <u>3</u> 4 5 6 7 8 9 10

10) 102

<u>2</u> <u>3</u> 4 5 <u>6</u> 7 8 9 10

11) 144

<u>2</u> <u>3</u> <u>4</u> 5 <u>6</u> 7 <u>8</u> <u>9</u> 10

12) 75

2 <u>3</u> 4 <u>5</u> 6 7 8 9 10

Chapter 3: Real Numbers and Integers

Topics that you'll learn in this chapter:

- ✓ Adding and Subtracting Integers
- ✓ Multiplying and Dividing Integers
- ✓ Ordering Integers and Numbers
- ✓ Arrange and Order, Comparing Integers
- ✓ Order of Operations
- ✓ Mixed Integer Computations
- ✓ Integers and Absolute Value

"Wherever there is number, there is beauty." –Proclus

Adding and Subtracting Integers

Helpful	-	**Integers:** {… , –3, –2, –1, 0, 1, 2, 3, …} Includes: zero, counting numbers, and the negative of the counting numbers.	**Example:**
Hints		– Add a positive integer by moving to the right on the number line.	$12 + 10 = 22$ $25 - 13 = 12$
		– Add a negative integer by moving to the left on the number line.	$(-24) + 12 = -12$ $(-14) + (-12) = -26$
		– Subtract an integer by adding its opposite.	$14 - (-13) = 27$

✎ *Find the sum.*

1) $(-12) + (-4)$

2) $5 + (-24)$

3) $(-14) + 23$

4) $(-8) + (39)$

5) $43 + (-12)$

6) $(-23) + (-4) + 3$

7) $4 + (-12) + (-10) + (-25)$

8) $19 + (-15) + 25 + 11$

9) $(-9) + (-12) + (32 - 14)$

10) $4 + (-30) + (45 - 34)$

✎ *Find the difference.*

11) $(-14) - (-9) - (18)$

12) $(-9) - (-25)$

13) $(-12) - (8)$

14) $(28) - (-4)$

15) $(34) - (2)$

16) $(55) - (-5) + (-4)$

17) $(9) - (2) - (-5)$

18) $(2) - (4) - (-15)$

19) $(23) - (4) - (-34)$

20) $(-45) - (-87)$

Multiplying and Dividing Integers

Helpful	(negative) × (negative) = positive	Examples:
	(negative) ÷ (negative) = positive	$3 \times 2 = 6$
Hints	(negative) × (positive) = negative	$3 \times -3 = -9$
	(negative) ÷ (positive) = negative	$-2 \times -2 = 4$
	(positive) × (positive) = positive	$10 \div 2 = 5$
		$-4 \div 2 = -2$
		$-12 \div -6 = 3$

✎ *Find each product.*

1) $(-8) \times (-2)$

2) 3×6

3) $(-4) \times 5 \times (-6)$

4) $2 \times (-6) \times (-6)$

5) $11 \times (-12)$

6) $10 \times (-5)$

7) 8×8

8) $(-8) \times (-9)$

9) $6 \times (-5) \times 3$

10) $6 \times (-1) \times 2$

✎ *Find each quotient.*

11) $18 \div 3$

12) $(-24) \div 4$

13) $(-63) \div (-9)$

14) $54 \div 9$

15) $20 \div (-2)$

16) $(-66) \div (-11)$

17) $64 \div 8$

18) $(-121) \div 11$

19) $72 \div 9$

20) $16 \div 4$

Ordering Integers and Numbers

Helpful	To compare numbers, you can use number line! As you move from left to right on the number line, you find a bigger number!	**Example:**
Hints		Order integers from least to greatest.
		$(-11, -13, 7, -2, 12)$
		$-13 < -11 < -2 < 7 < 12$

✎ *Order each set of integers from least to greatest.*

1) $-15, -19, 20, -4, 1$ ___, ___, ___, ___, ___, ___

2) $6, -5, 4, -3, 2$ ___, ___, ___, ___, ___, ___

3) $15, -42, 19, 0, -22$ ___, ___, ___, ___, ___

4) $26, -91, 0, -13, 67, -55$ ___, ___, ___, ___, ___, ___

5) $-17, -71, 90, -25, -54, -39$ ___, ___, ___, ___, ___, ___

6) $98, 5, 46, 19, 77, 24$ ___, ___, ___, ___, ___, ___

✎ *Order each set of integers from greatest to least.*

7) $-2, 5, -3, 6, -4$ ___, ___, ___, ___, ___, ___

8) $-37, 7, -17, 27, 47$ ___, ___, ___, ___, ___, ___

9) $32, -27, 19, -17, 15$ ___, ___, ___, ___, ___, ___

10) $68, 81, 21, -18, 94, 72$ ___, ___, ___, ___, ___, ___

Arrange, Order, and Comparing Integers

Helpful	When using a number line, numbers increase as you move to the right.	**Examples:**
Hints		$5 < 7$,
		$-5 < -2$
		$-18 < -12$

✍ *Arrange these integers in descending order.*

1) $21, 71, -18, -10, 82$ ___, ___, ___, ___, ___, ___

2) $15, 11, 20, 12, -9, -5$ ___, ___, ___, ___, ___, ___

3) $-5, 20, 15, 9, -11$ ___, ___, ___, ___, ___, ___

4) $19, 18, -9, -6, -11$ ___, ___, ___, ___, ___, ___

5) $56, -34, -12, -5, 32$ ___, ___, ___, ___, ___, ___

✍ *Compare. Use >, =, <*

6) -8 ____ 12 11) -56 ____ -58

7) -10 ____ -16 12) 78 ____ 87

8) 43 ____ 34 13) -92 ____ -102

9) 15 ____ -16 14) -12 ____ -12

10) -354 ____ -345 15) -721 ____ -821

Order of Operations

Helpful	-	Use "order of operations" rule when there are more than one math operation.	Example:
Hints	-	PEMDAS (parentheses / exponents / multiply / divide / add / subtract)	$(12 + 4) \div (-4) = -4$

✎ *Evaluate each expression.*

1) $(2 \times 2) + 5$

2) $24 - (3 \times 3)$

3) $(6 \times 4) + 8$

4) $25 - (4 \times 2)$

5) $(6 \times 5) + 3$

6) $64 - (2 \times 4)$

7) $25 + (1 \times 8)$

8) $(6 \times 7) + 7$

9) $48 \div (4 + 4)$

10) $(7 + 11) \div (-2)$

11) $9 + (2 \times 5) + 10$

12) $(5 + 8) \times \frac{3}{5} + 2$

13) $2 \times 7 - (\frac{10}{9 - 4})$

14) $(12 + 2 - 5) \times 7 - 1$

15) $(\frac{7}{5 - 1}) \times (2 + 6) \times 2$

16) $20 \div (4 - (10 - 8))$

17) $\frac{50}{4(5 - 4) - 3}$

18) $2 + (8 \times 2)$

Mixed Integer Computations

Helpful Hints	It worth remembering:	Example:
	(negative) × (negative) = positive	
	(negative) ÷ (negative) = positive	(−5) + 6 = 1
	(negative) × (positive) = negative	(−3) × (−2) = 6
	(negative) ÷ (positive) = negative	(9) ÷ (−3) = − 3
	(positive) × (positive) = positive	

✎ **Compute.**

1) $(-70) \div (-5)$

2) $(-14) \times 3$

3) $(-4) \times (-15)$

4) $(-65) \div 5$

5) $18 \times (-7)$

6) $(-12) \times (-2)$

7) $\dfrac{(-60)}{(-20)}$

8) $24 \div (-8)$

9) $22 \div (-11)$

10) $\dfrac{(-27)}{3}$

11) $4 \times (-4)$

12) $\dfrac{(-48)}{12}$

13) $(-14) \times (-2)$

14) $(-7) \times (7)$

15) $\dfrac{-30}{-6}$

16) $(-54) \div 6$

17) $(-60) \div (-5)$

18) $(-7) \times (-12)$

19) $(-14) \times 5$

20) $88 \div (-8)$

Integers and Absolute Value

Helpful Hints	To find an absolute value of a number, just find it's distance from 0!	Example: $\|-6\| = 6$ $\|6\| = 6$ $\|-12\| = 12$ $\|12\| = 12$

✎ *Write absolute value of each number.*

1) $- 4$

2) $- 7$

3) $- 8$

4) 4

5) 5

6) $- 10$

7) 1

8) 6

9) 8

10) $- 2$

11) $- 1$

12) 10

13) 3

14) 7

15) $- 5$

16) $- 3$

17) $- 9$

18) 2

19) 4

20) $- 6$

21) 9

✎ *Evaluate.*

22) $\|-43\| - \|12\| + 10$

23) $76 + \|-15 - 45\| - \|3\|$

24) $30 + \|-62\| - 46$

25) $\|32\| - \|-78\| + 90$

26) $\|-35 + 4\| + 6 - 4$

27) $\|-4\| + \|-11\|$

28) $\|-6 + 3 - 4\| + \|7 + 7\|$

29) $\|-9\| + \|-19\| - 5$

Answers of Worksheets – CHAPTER 3

Adding and Subtracting Integers

1) -16	8) 40	15) 32
2) -19	9) -3	16) 56
3) 9	10) -15	17) 12
4) 31	11) -23	18) 13
5) 31	12) 16	19) 53
6) -24	13) -20	20) 42
7) -43	14) 32	

Multiplying and Dividing Integers

1) 16	8) 72	15) -10
2) 18	9) -90	16) 6
3) 120	10) -12	17) 8
4) 72	11) 6	18) -11
5) -132	12) -6	19) 8
6) -50	13) 7	20) 4
7) 64	14) 6	

Ordering Integers and Numbers

1) $-19, -15, -4, 1, 20$	6) $5, 19, 24, 46, 77, 98$
2) $-5, -3, 2, 4, 6$	7) $6, 5, -2, -3, -4$
3) $-42, -22, 0, 15, 19$	8) $47, 27, 7, -17, -37$
4) $-91, -55, -13, 0, 26, 67$	9) $32, 19, 15, -17, -27$
5) $-71, -54, -39, -25, -17, 90$	10) $94, 81, 72, 68, 21, -18$

Arrange and Order, Comparing Integers

1) $82, 71, 21, -10, -18$

2) $20, 15, 12, 11, -5, -9$

3) $20, 15, 9, -5, -11$

4) $19, 18, -6, -9, -11$

5) $56, 32, -5, -12, -34$

6) $<$ 10) $<$ 14) $=$

7) $>$ 11) $>$ 15) $>$

8) $>$ 12) $<$

9) $>$ 13) $>$

Order of Operations

1) 9   7) 33   13) 12

2) 15   8) 49   14) 62

3) 32   9) 6   15) 28

4) 17   10) -9 16) 10

5) 33   11) 29   17) 50

6) 56   12) 9.8   18) 18

Mixed Integer Computations

1) 14   8) -3 15) 5

2) -42 9) -2 16) -9

3) 60   10) -9 17) 12

4) -13 11) -16 18) 84

5) -126 12) -4 19) -70

6) 24   13) 28   20) -11

7) 3   14) -49

Integers and Absolute Value

1) 4	11) 1	21) 9
2) 7	12) 10	22) 41
3) 8	13) 3	23) 133
4) 4	14) 7	24) 46
5) 5	15) 5	25) 44
6) 10	16) 3	26) 33
7) 1	17) 9	27) 15
8) 6	18) 2	28) 21
9) 8	19) 4	29) 23
10) 2	20) 6	

Chapter 4: Proportions and Ratios

Topics that you'll learn in this chapter:

- ✓ Writing Ratios
- ✓ Simplifying Ratios
- ✓ Create a Proportion
- ✓ Similar Figures
- ✓ Simple Interest
- ✓ Ratio and Rates Word Problems

"Do not worry about your difficulties in mathematics. I can assure you mine are still greater."

– Albert Einstein

Writing Ratios

Helpful	– A ratio is a comparison of two numbers. Ratio can be written as a division.	Example:
Hints		$3 : 5,$ or $\frac{3}{5}$

✍ *Express each ratio as a rate and unite rate.*

1) 120 miles on 4 gallons of gas.

2) 24 dollars for 6 books.

3) 200 miles on 14 gallons of gas

4) 24 inches of snow in 8 hours

✍ *Express each ratio as a fraction in the simplest form.*

5) 3 feet out of 30 feet

6) 18 cakes out of 42 cakes

7) 16 dimes t0 24 dimes

8) 12 dimes out of 48 coins

9) 14 cups to 84 cups

10) 45 gallons to 65 gallons

11) 10 miles out of 40 miles

12) 22 blue cars out of 55 cars

13) 32 pennies to 300 pennies

14) 24 beetles out of 86 insects

Simplifying Ratios

Helpful *Hints*	− You can calculate equivalent ratios by multiplying or dividing both sides of the ratio by the same number.	**Examples:** $3 : 6 = 1 : 2$ $4 : 9 = 8 : 18$

✎*Reduce each ratio.*

1) $21 : 49$

2) $20 : 40$

3) $10 : 50$

4) $14 : 18$

5) $45 : 27$

6) $49 : 21$

7) $100 : 10$

8) $12 : 8$

9) $35 : 45$

10) $8 : 20$

11) $25 : 35$

12) $21 : 27$

13) $52 : 82$

14) $12 : 36$

15) $24 : 3$

16) $15 : 30$

17) $3 : 36$

18) $8 : 16$

19) $6 : 100$

20) $2 : 20$

21) $10 : 60$

22) $14 : 63$

23) $68 : 80$

24) $8 : 80$

Create a Proportion

Helpful	— A proportion contains 2 equal fractions! A proportion simply means that two fractions are equal.	**Example:**
Hints		2, 4, 8, 16 $$\frac{2}{4} = \frac{8}{16}$$

✍️ *Create proportion from the given set of numbers.*

1) 1, 6, 2, 3   7) 10, 5, 8, 4

2) 12, 144, 1, 12   8) 3, 12, 8, 2

3) 16, 4, 8, 2   9) 2, 2, 1, 4

4) 9, 5, 27, 15   10)   3, 6, 7, 14

5) 7, 10, 60, 42   11)   2, 6, 5, 15

6) 8, 7, 24, 21   12)   7, 2, 14, 4

Similar Figures

Helpful	– Two or more figures are similar if the corresponding angles are equal, and the corresponding sides are in proportion.	**Example:**
Hints		3–4–5 triangle is similar to a 6–8–10 triangle

✎ *Each pair of figures is similar. Find the missing side.*

1)

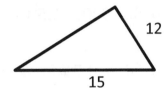 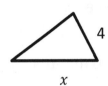

2)

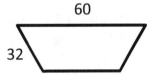

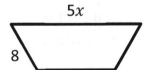 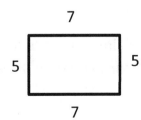

3)

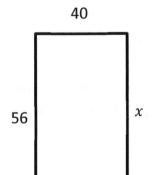

Simple Interest

Helpful Hints	**Simple Interest:** The charge for borrowing money or the return for lending it. Interest = principal x rate x time $$I = prt$$	**Example:** $450 at 7% for 8 years. $$I = prt$$ $$I = 450 \times 0.07 \times 8 = \$252$$

✏️ **Use simple interest to find the ending balance.**

1) $1,300 at 5% for 6 years.

2) $5,400 at 7.5% for 6 months.

3) $25,600 at 9.2% for 5 years

4) $24,000 at 8.5% for 9 years.

5) $450 at 7% for 8 years.

6) $54,200 at 8% for 5 years.

7) $240 interest is earned on a principal of $1500 at a simple interest rate of 4% interest per year. For how many years was the principal invested?

8) A new car, valued at $28,000, depreciates at 9% per year from original price. Find the value of the car 3 years after purchase.

9) Sara puts $2,000 into an investment yielding 5% annual simple interest; she left the money in for five years. How much interest does Sara get at the end of those five years?

Ratio and Rates Word Problems

Helpful Hints	To solve a ratio or a rate word problem, create a proportion and use cross multiplication method!	**Example:** $$\frac{x}{4} = \frac{8}{16}$$ $$16x = 4 \times 8$$ $$x = 2$$

✎ Solve.

1) In a party, 10 soft drinks are required for every 12 guests. If there are 252 guests, how many soft drinks is required?

2) In Jack's class, 18 of the students are tall and 10 are short. In Michael's class 54 students are tall and 30 students are short. Which class has a higher ratio of tall to short students?

3) Are these ratios equivalent?

   12 cards to 72 animals, 11 marbles to 66 marbles

4) The price of 3 apples at the Quick Market is $1.44. The price of 5 of the same apples at Walmart is $2.50. Which place is the better buy?

5) The bakers at a Bakery can make 160 bagels in 4 hours. How many bagels can they bake in 16 hours? What is that rate per hour?

6) You can buy 5 cans of green beans at a supermarket for $3.40. How much does it cost to buy 35 cans of green beans?

Answers of Worksheets – Chapter 4

Writing Ratios

1) $\frac{120\ miles}{4\ gallons}$, 30 miles per gallon

2) $\frac{24\ dollars}{6\ books}$, 4.00 dollars per book

3) $\frac{200\ miles}{14\ gallons}$, 14.29 miles per gallon

4) $\frac{24"\ of\ snow}{8\ hou}$, 3 inches of snow per hour

5) $\frac{1}{10}$

6) $\frac{3}{7}$

7) $\frac{2}{3}$

8) $\frac{1}{4}$

9) $\frac{1}{6}$

10) $\frac{9}{13}$

11) $\frac{1}{4}$

12) $\frac{2}{5}$

13) $\frac{8}{75}$

14) $\frac{12}{43}$

Simplifying Ratios

1) 3 : 7
2) 1 : 2
3) 1 : 5
4) 7 : 9
5) 5 : 3
6) 7 : 3
7) 10 : 1
8) 3 : 2

9) 7 : 9
10) 2 : 5
11) 5 : 7
12) 7 : 9
13) 26 : 41
14) 1 : 3
15) 8 : 1
16) 1 : 2

17) 1 : 12
18) 1 : 2
19) 3 : 50
20) 1 : 10
21) 1: 6
22) 2 : 9
23) 17 : 20
24) 1 : 10

Create a Proportion

1) 1 : 3 = 2 : 6
2) 12 : 144 = 1 : 12
3) 2 : 4 = 8 : 16

4) 5 : 15 = 9 : 27
5) 7 : 42, 10 : 60
6) 7 : 21 = 8 : 24

7) 8 : 10 = 4 : 5
8) 2 : 3 = 8 : 12
9) 4 : 2 = 2 : 1

10) 7 : 3 = 14 : 6   11) 5 : 2 = 15 : 6   12) 7 : 2 = 14 : 4

Similar Figures

1) 5   2) 3   3) 56

Simple Interest

1) $1,690.00   4) $42,360.00   7) 4 years
2) $5,602.50   5) $702.00   8) $20,440
3) $37,376.00   6) $75,880.00   9) $500

Ratio and Rates Word Problems

1) 210

2) The ratio for both class is equal to 9 to 5.

3) Yes! Both ratios are 1 to 6

4) The price at the Quick Market is a better buy.

5) 640, the rate is 40 per hour.

6) $23.80

Chapter 5: Percent

Topics that you'll learn in this chapter:

- ✓ Percentage Calculations
- ✓ Converting Between Percent, Fractions, and Decimals
- ✓ Percent Problems
- ✓ Markup, Discount, and Tax

"The book of nature is written in the language of Mathematic" -Galileo

Percentage Calculations

Helpful Hints	-	Use the following formula to find part, whole, or percent: $\text{part} = \dfrac{\text{percent}}{100} \times \text{whole}$	**Example:** $\dfrac{20}{100} \times 100 = 20$

✎ **Calculate the percentages.**

1) 50% of 25

2) 80% of 15

3) 30% of 34

4) 70% of 45

5) 10% of 0

6) 80% of 22

7) 65% of 8

8) 78% of 54

9) 50% of 80

10) 20% of 10

11) 40% of 40

12) 90% of 0

13) 20% of 70

14) 55% of 60

15) 80% of 10

16) 20% of 880

17) 70% of 100

18) 80% of 90

✎ **Solve.**

19) 50 is what percentage of 75?

20) What percentage of 100 is 70

21) Find what percentage of 60 is 35.

22) 40 is what percentage of 80?

Converting Between Percent, Fractions, and Decimals

Helpful Hints	– To a percent: Move the decimal point 2 places to the right and add the % symbol.	**Examples:**
	– Divide by 100 to convert a number from percent to decimal.	30% = 0.3
		0.24 = 24%

✐ *Converting fractions to decimals.*

1) $\dfrac{50}{100}$ 4) $\dfrac{80}{100}$ 7) $\dfrac{90}{100}$

2) $\dfrac{38}{100}$ 5) $\dfrac{7}{100}$ 8) $\dfrac{20}{100}$

3) $\dfrac{15}{100}$ 6) $\dfrac{35}{100}$ 9) $\dfrac{7}{100}$

✐ *Write each decimal as a percent.*

10) 0.5   13) 0.524   16) 3.63

11) 0.9   14) 0.1   17) 0.008

12) 0.002   15) 0.03   18) 4.78

Percent Problems

Helpful	Base = Part ÷ Percent	**Example:**
	Part = Percent × Base	2 is 10% of 20.
Hints	Percent = Part ÷ Base	$2 ÷ 0.10 = 20$
		$2 = 0.10 × 20$
		$0.10 = 2 ÷ 20$

✍ *Solve each problem.*

1) 51 is 340% of what?

2) 93% of what number is 97?

3) 27% of 142 is what number?

4) What percent of 125 is 29.3?

5) 60 is what percent of 126?

6) 67 is 67% of what?

7) 67 is 13% of what?

8) 41% of 78 is what?

9) 1 is what percent of 52.6?

10) What is 59% of 14 m?

11) What is 90% of 130 inches?

12) 16 inches is 35% of what?

13) 90% of 54.4 hours is what?

14) What percent of 33.5 is 21?

15) Liam scored 22 out of 30 marks in Algebra, 35 out of 40 marks in science and 89 out of 100 marks in mathematics. In which subject his percentage of marks in best?

16) Ella require 50% to pass. If she gets 280 marks and falls short by 20 marks, what were the maximum marks she could have got?

Markup, Discount, and Tax

		Example:
Helpful	- **Markup** = selling price − cost Markup rate = markup divided by the cost	
Hints	- **Discount:** Multiply the regular price by the rate of discount Selling price = original price − discount	Original price of a microphone: $49.99, discount: 5%, tax: 5%
	- **Tax:** To find tax, multiply the tax rate to the taxable amount (income, property value, etc.)	*Selling price = 49.87*

✎ *Find the selling price of each item.*

1) Cost of a pen: $1.95, markup: 70%, discount: 40%, tax: 5%

2) Cost of a puppy: $349.99, markup: 41%, discount: 23%

3) Cost of a shirt: $14.95, markup: 25%, discount: 45%

4) Cost of an oil change: $21.95, markup: 95%

5) Cost of computer: $1,850.00, markup: 75%

Answers of Worksheets – Chapter 5

Percentage Calculations

1) 12.5	9) 40	17) 70
2) 12	10) 2	18) 72
3) 10.2	11) 16	19) 67%
4) 31.5	12) 0	20) 70%
5) 0	13) 14	21) 58%
6) 17.6	14) 33	22) 50%
7) 5.2	15) 8	
8) 42.12	16) 176	

Converting Between Percent, Fractions, and Decimals

1) 0.5	7) 0.9	13) 52.4%
2) 0.38	8) 0.2	14) 10%
3) 0.15	9) 0.07	15) 3%
4) 0.8	10) 50%	16) 363%
5) 0.07	11) 90%	17) 0.8%
6) 0.35	12) 0.2%	18) 478%

Percent Problems

1) 15	7) 515.4	13) 49 hours
2) 104.3	8) 31.98	14) 62.7%
3) 38.34	9) 1.9%	15) Mathematics
4) 23.44%	10) 8.3 m	16) 600
5) 47.6%	11) 117 inches	
6) 100	12) 45.7inches	

Markup, Discount, and Tax

1) $2.09

2) $379.98

3) $10.28

4) $36.22

5) $3,237.50

Chapter 6: Algebraic Expressions

Topics that you'll learn in this chapter:

- ✓ Expressions and Variables
- ✓ Simplifying Variable Expressions
- ✓ Simplifying Polynomial Expressions
- ✓ Translate Phrases into an Algebraic Statement
- ✓ The Distributive Property
- ✓ Evaluating One Variable
- ✓ Evaluating Two Variables
- ✓ Combining like Terms

Without mathematics, there's nothing you can do. Everything around you is mathematics. Everything around you is numbers." – Shakuntala Devi

Expressions and Variables

Helpful	A variable is a letter that represents unknown numbers. A variable can be used in the same manner as all other numbers:

Hints			
	Addition	$2 + a$	2 plus a
	Subtraction	$y - 3$	y minus 3
	Division	$\dfrac{4}{x}$	4 divided by x
	Multiplication	$5a$	5 times a

✍ *Simplify each expression.*

1) $x + 5x$,

   use $x = 5$

2) $8(-3x + 9) + 6$,

   use $x = 6$

3) $10x - 2x + 6 - 5$,

   use $x = 5$

4) $2x - 3x - 9$,

   use $x = 7$

5) $(-6)(-2x - 4y)$,

   use $x = 1$, $y = 3$

6) $8x + 2 + 4y$,

   use $x = 9$, $y = 2$

7) $(-6)(-8x - 9y)$,

   use $x = 5$, $y = 5$

8) $6x + 5y$,

   use $x = 7$, $y = 4$

✍ *Simplify each expression.*

9) $5(-4 + 2x)$

10) $-3 - 5x - 6x + 9$

11) $6x - 3x - 8 + 10$

12) $(-8)(6x - 4) + 12$

13) $9(7x + 4) + 6x$

14) $(-9)(-5x + 2)$

Simplifying Variable Expressions

Helpful	– Combine "like" terms. (values with same variable and same power)	**Example:**
Hints	– Use distributive property if necessary.	$2x + 2\,(1 - 5x) =$
		$2x + 2 - 10x = -8x + 2$
	Distributive Property:	
	$a\,(b\,+\,c)\,=\,ab\,+\,ac$	

✍ *Simplify each expression.*

1) $-2 - x^2 - 6x^2$

2) $3 + 10x^2 + 2$

3) $8x^2 + 6x + 7x^2$

4) $5x^2 - 12x^2 + 8x$

5) $2x^2 - 2x - x$

6) $(-6)\,(8x - 4)$

7) $4x + 6\,(2 - 5x)$

8) $10x + 8\,(10x - 6)$

9) $9\,(-2x - 6) - 5$

10) $3\,(x + 9)$

11) $7x + 3 - 3x$

12) $2.5x^2 \times (-8x)$

✍ *Simplify.*

13) $-2(4 - 6x) - 3x,\ x = 1$

14) $2x + 8x,\ x = 2$

15) $9 - 2x + 5x + 2,\ x = 5$

16) $5\,(3x + 7),\ x = 3$

17) $2\,(3 - 2x) - 4,\ x = 6$

18) $5x + 3x - 8,\ x = 3$

19) $x - 7x,\ x = 8$

20) $5\,(-2 - 9x),\ x = 4$

Simplifying Polynomial Expressions

Helpful	-	In mathematics, a polynomial is an expression consisting of variables and coefficients that involves only the operations of addition, subtraction, multiplication, and non–negative integer exponents of variables. $P(x) = a_0 x^n + a_1 x^{n-1} + ... + a_{n-2} 2x^2 + a_{n-1}x + a_n$	**Example:** An example of a polynomial of a single indeterminate x is $x^2 - 4x + 7$. An example for three variables is $x^3 + 2xyz^2 - yz + 1$
Hints			

✍ *Simplify each polynomial.*

1) $4x^5 - 5x^6 + 15x^5 - 12x^6 + 3x^6$

2) $(-3x^5 + 12 - 4x) + (8x^4 + 5x + 5x^5)$

3) $10x^2 - 5x^4 + 14x^3 - 20x^4 + 15x^3 - 8x^4$

4) $-6x^2 + 5x^2 - 7x^3 + 12 + 22$

5) $12x^5 - 5x^3 + 8x^2 - 8x^5$

6) $5x^3 + 1 + x^2 - 2x - 10x$

7) $14x^2 - 6x^3 - 2x(4x^2 + 2x)$

8) $(4x^4 - 2x) - (4x - 2x^4)$

9) $(3x^2 + 1) - (4 + 2x^2)$

10) $(2x + 2) - (7x + 6)$

11) $(12x^3 + 4x^4) - (2x^4 - 6x^3)$

12) $(12 + 3x^3) + (6x^3 + 6)$

13) $(5x^2 - 3) + (2x^2 - 3x^3)$

14) $(23x^3 - 12x^2) - (2x^2 - 9x^3)$

15) $(4x - 3x^3) - (3x^3 + 4x)$

Translate Phrases into an Algebraic Statement

Helpful	**Translating key words and phrases into algebraic expressions:**
	Addition: plus, more than, the sum of, etc.
Hints	**Subtraction:** minus, less than, decreased, etc.
	Multiplication: times, product, multiplied, etc.
	Division: quotient, divided, ratio, etc.
	Example:
	eight more than a number is 20
	$8 + x = 20$

✍ *Write an algebraic expression for each phrase.*

1) A number increased by forty–two.

2) The sum of fifteen and a number

3) The difference between fifty–six and a number.

4) The quotient of thirty and a number.

5) Twice a number decreased by 25.

6) Four times the sum of a number and – 12.

7) A number divided by – 20.

8) The quotient of 60 and the product of a number and – 5.

9) Ten subtracted from a number.

10) The difference of six and a number.

The Distributive Property

Helpful	Distributive Property:	Example:
Hints	$a\,(b\,+\,c)\,=\,ab\,+\,ac$	$3\,(4\,+\,3x)$
		$=\,12\,+\,9x$

✎ **Use the distributive property to simply each expression.**

1) $-(-2-5x)$ 6) $2\,(12+2x)$

2) $(-6x+2)(-1)$ 7) $(-6x+8)\,4$

3) $(-5)\,(x-2)$ 8) $(3-6x)(-7)$

4) $-(7-3x)$ 9) $(-12)\,(2x+1)$

5) $8\,(8+2x)$ 10) $(8-2x)\,9$

11) $(-2x)\,(-1+9x)-4x\,(4+5x)$

12) $3\,(-5x-3)+4(6-3x)$

13) $(-2)(x+4)-(2+3x)$

14) $(-4)(3x-2)+6\,(x+1)$

15) $(-5)(4x-1)+4\,(x+2)$

16) $(-3)(x+4)-(2+3x)$

Evaluating One Variable

Helpful *Hints*	– To evaluate one variable expression, find the variable and substitute a number for that variable. – Perform the arithmetic operations.	**Example:** $4x + 8, x = 6$ $4(6) + 8 = 24 + 8 = 32$

✍ *Simplify each algebraic expression.*

1) $9 - x$, $x = 3$

2) $x + 2, x = 5$

3) $3x + 7, x = 6$

4) $x + (-5), x = -2$

5) $3x + 6, x = 4$

6) $4x + 6, x = -1$

7) $10 + 2x - 6, x = 3$

8) $10 - 3x, x = 8$

9) $\dfrac{20}{x} - 3, x = 5$

10) $(-3) + \dfrac{x}{4} + 2x, x = 16$

11) $(-2) + \dfrac{x}{7}, x = 21$

12) $(-\dfrac{14}{x}) - 9 + 4x, x = 2$

13) $(-\dfrac{6}{x}) - 9 + 2x, x = 3$

14) $(-2) + \dfrac{x}{8}, x = 16$

Evaluating Two Variables

Helpful	To evaluate an algebraic expression, substitute a number for each variable and perform the arithmetic operations.	**Example:**
Hints		$2x + 4y - 3 + 2,$
		$x = 5, y = 3$
		$2(5) + 4(3) - 3 + 2$
		$= 10$
		$+ 12 - 3 + 2$
		$= 21$

✎**Simplify each algebraic expression.**

1) $2x + 4y - 3 + 2,$

   $x = 5, y = 3$

2) $(-\frac{12}{x}) + 1 + 5y,$

   $x = 6, y = 8$

3) $(-4)(-2a - 2b),$

   $a = 5, b = 3$

4) $10 + 3x + 7 - 2y,$

   $x = 7, y = 6$

5) $9x + 2 - 4y,$

   $x = 7, y = 5$

6) $6 + 3(-2x - 3y),$

   $x = 9, y = 7$

7) $12x + y,$

   $x = 4, y = 8$

8) $x \times 4 \div y,$

   $x = 3, y = 2$

9) $2x + 14 + 4y,$

   $x = 6, y = 8$

10) $4a - (5 - b),$

    $a = 4, b = 6$

Combining like Terms

Helpful	– Terms are separated by "+" and "–" signs.	**Example:**
	– Like terms are terms with same variables and same powers.	$22x + 6 + 2x =$
Hints	– Be sure to use the "+" or "–" that is in front of the coefficient.	$24x + 6$

✎ *Simplify each expression.*

1) $5 + 2x - 8$

2) $(-2x + 6)\, 2$

3) $7 + 3x + 6x - 4$

4) $(-4) - (3)(5x + 8)$

5) $9x - 7x - 5$

6) $x - 12x$

7) $7\,(3x + 6) + 2x$

8) $(-11x) - 10x$

9) $3x - 12 - 5x$

10) $13 + 4x - 5$

11) $(-22x) + 8x$

12) $2\,(4 + 3x) - 7x$

13) $(-4x) - (6 - 14x)$

14) $5\,(6x - 1) + 12x$

15) $22x + 6 + 2x$

16) $(-13x) - 14x$

17) $(-6x) - 9 + 15x$

18) $(-6x) + 7x$

19) $(-5x) + 12 + 7x$

20) $(-3x) - 9 + 15x$

21) $20x - 19x$

Answers of Worksheets – Chapter 6

Expressions and Variables

1) 30
2) −66
3) 41
4) −16
5) 84
6) 82
7) 510
8) 62
9) $10x - 20$
10) $6 - 11x$
11) $3x + 2$
12) $44 - 48x$
13) $69x + 36$
14) $45x - 18$

Simplifying Variable Expressions

1) $-7x^2 - 2$
2) $10x^2 + 5$
3) $15x^2 + 6x$
4) $-7x^2 + 8x$
5) $2x^2 - 3x$
6) $-48x + 24$
7) $-26x + 12$
8) $90x - 48$
9) $-18x - 59$
10) $3x + 27$
11) $4x + 3$
12) $-20x^3$
13) 1
14) 20
15) 26
16) 80
17) -22
18) 16
19) -48
20) -190

Simplifying Polynomial Expressions

1) $-14x^6 + 19x^5$
2) $2x^5 + 8x^4 + x + 12$
3) $-33x^4 + 29x^3 + 10x^2$
4) $-7x^3 - x^2 + 34$
5) $4x^5 - 5x^3 + 8x^2$
6) $5x^3 + x^2 - 12x + 1$
7) $-14x^3 + 10x^2$
8) $6x^4 - 6x$
9) $x^2 - 3$
10) $-5x - 4$
11) $2x^4 + 18x^3$
12) $9x^3 + 18$
13) $-3x^3 + 7x^2 - 3$
14) $32x^3 - 14x^2$
15) $-6x^3$

Translate Phrases into an Algebraic Statement

1) $x + 42$
2) $15 + x$
3) $56 - x$
4) $30/x$
5) $2x - 25$
6) $4(x + (-12))$
7) $\dfrac{x}{-20}$
8) $\dfrac{60}{-5x}$
9) $x - 10$
10) $6 - x$

The Distributive Property

1) $5x + 2$
2) $6x - 2$
3) $-5x + 10$
4) $3x - 7$
5) $16x + 64$
6) $4x + 24$

7) $-24x + 32$
8) $42x - 21$
9) $-24x - 12$
10) $-18x + 72$
11) $-38x^2 - 14x$
12) $-27x + 15$

13) $-5x - 10$
14) $-6x + 14$
15) $-16x + 13$
16) $-6x - 14$

Evaluating One Variable

1) 6
2) 7
3) 25
4) -7
5) 18

6) 2
7) 10
8) -14
9) 1
10) 33

11) 1
12) -8
13) -5
14) 0

Evaluating Two Variables

1) 21
2) 39
3) 64
4) 26

5) 45
6) -111
7) 56
8) 6

9) 58
10) 17

Combining like Terms

1) $2x - 3$
2) $-4x + 12$
3) $9x + 3$
4) $-15x - 28$
5) $2x - 5$
6) $-11x$
7) $23x + 42$

8) $-21x$
9) $-2x - 12$
10) $4x + 8$
11) $-14x$
12) $-x + 8$
13) $10x - 6$
14) $42x - 5$

15) $24x + 6$
16) $-27x$
17) $9x - 9$
18) x
19) $2x + 12$
20) $12x - 9$
21) x

Chapter 7: Equations

Topics that you'll learn in this chapter:

- ✓ One– Step Equations
- ✓ Two– Step Equations
- ✓ Multi– Step Equations

"The study of mathematics, like the Nile, begins in minuteness but ends in magnificence."

– Charles Caleb Colton

One–Step Equations

Helpful Hints	- The values of two expressions on both sides of an equation are equal. $$ax + b = c$$ - You only need to perform one Math operation in order to solve the equation.	**Example:** $-8x = 16$ $x = -2$

✎ **Solve each equation.**

1) $x + 3 = 17$

2) $22 = (-8) + x$

3) $3x = (-30)$

4) $(-36) = (-6x)$

5) $(-6) = 4 + x$

6) $2 + x = (-2)$

7) $20x = (-220)$

8) $18 = x + 5$

9) $(-23) + x = (-19)$

10) $5x = (-45)$

11) $x - 12 = (-25)$

12) $x - 3 = (-12)$

13) $(-35) = x - 27$

14) $8 = 2x$

15) $(-6x) = 36$

16) $(-55) = (-5x)$

17) $x - 30 = 20$

18) $8x = 32$

19) $36 = (-4x)$

20) $4x = 68$

21) $30x = 300$

Two–Step Equations

Helpful	– You only need to perform two math operations (add, subtract, multiply, or divide) to solve the equation.	**Example:**
Hints	– Simplify using the inverse of addition or subtraction.	$- 2 (x - 1) = 42$
	– Simplify further by using the inverse of multiplication or division.	$(x - 1) = - 21$
		$x = - 20$

✎ *Solve each equation.*

1) $5 (8 + x) = 20$

2) $(- 7) (x - 9) = 42$

3) $(- 12) (2x - 3) = (- 12)$

4) $6 (1 + x) = 12$

5) $12 (2x + 4) = 60$

6) $7 (3x + 2) = 42$

7) $8 (14 + 2x) = (- 34)$

8) $(- 15) (2x - 4) = 48$

9) $3 (x + 5) = 12$

10) $\dfrac{3x - 12}{6} = 4$

11) $(- 12) = \dfrac{x + 15}{6}$

12) $110 = (- 5)(2x - 6)$

13) $\dfrac{x}{8} - 12 = 4$

14) $20 = 12 + \dfrac{x}{4}$

15) $\dfrac{- 24 + x}{6} = (- 12)$

16) $(- 4) (5 + 2x) = (- 100)$

17) $(- 12x) + 20 = 32$

18) $\dfrac{-2 + 6x}{4} = (- 8)$

19) $\dfrac{x + 6}{5} = (- 5)$

20) $(- 9) + \dfrac{x}{4} = (- 15)$

Multi–Step Equations

Helpful *Hints*	– Combine "like" terms on one side. – Bring variables to one side by adding or subtracting. – Simplify using the inverse of addition or subtraction. – Simplify further by using the inverse of multiplication or division.	**Example:** $3x + 15 = -2x + 5$ Add 2x both sides $5x + 15 = +5$ Subtract 15 both sides $5x = -10$ Divide by 5 both sides $x = -2$

✎ *Solve each equation.*

1) $-(2 - 2x) = 10$

2) $-12 = -(2x + 8)$

3) $3x + 15 = (-2x) + 5$

4) $-28 = (-2x) - 12x$

5) $2(1 + 2x) + 2x = -118$

6) $3x - 18 = 22 + x - 3 + x$

7) $12 - 2x = (-32) - x + x$

8) $7 - 3x - 3x = 3 - 3x$

9) $6 + 10x + 3x = (-30) + 4x$

10) $(-3x) - 8(-1 + 5x) = 352$

11) $24 = (-4x) - 8 + 8$

12) $9 = 2x - 7 + 6x$

13) $6(1 + 6x) = 294$

14) $-10 = (-4x) - 6x$

15) $4x - 2 = (-7) + 5x$

16) $5x - 14 = 8x + 4$

17) $40 = -(4x - 8)$

18) $(-18) - 6x = 6(1 + 3x)$

19) $x - 5 = -2(6 + 3x)$

20) $6 = 1 - 2x + 5$

Answers of Worksheets – Chapter 7

One–Step Equations

1) 14

2) 30

3) − 10

4) 6

5) − 10

6) − 4

7) − 11

8) 13

9) 4

10) − 9

11) − 13

12) − 9

13) − 8

14) 4

15) − 6

16) 11

17) 50

18) 4

19) − 9

20) 17

21) 10

Two–Step Equations

1) − 4

2) 3

3) 2

4) 1

5) 0.5

6) $\frac{4}{3}$

7) $-\frac{73}{8}$

8) $\frac{2}{5}$

9) − 1

10) 12

11) − 87

12) − 8

13) 128

14) 32

15) − 48

16) 10

17) − 1

18) − 5

19) − 31

20) − 24

Multi–Step Equations

1) 6

2) 2

3) − 2

4) 2

5) − 20

6) 37

7) 22

8) $\frac{4}{3}$

9) − 4

10) − 8

11) − 6

12) 2

13) 8

14) 1

15) 5

16) − 6

17) − 8

18) − 1

19) − 1

20) 0

Chapter 8: Inequalities

Topics that you'll learn in this chapter:

- ✓ Graphing Single– Variable Inequalities
- ✓ One– Step Inequalities
- ✓ Two– Step Inequalities
- ✓ Multi– Step Inequalities

Without mathematics, there's nothing you can do. Everything around you is mathematics. Everything around you is numbers." – Shakuntala Devi

Graphing Single–Variable Inequalities

Helpful	– Isolate the variable.
	– Find the value of the inequality on the number line.
Hints	– For less than or greater than draw open circle on the value of the variable.
	– If there is an equal sign too, then use filled circle.
	– Draw a line to the right direction.

✎ *Draw a graph for each inequality.*

1) $-2 > x$

2) $5 \leq -x$

3) $x > 7$

4) $-x > 1.5$

One–Step Inequalities

Helpful *Hints*	– Isolate the variable. – For dividing both sides by negative numbers, flip the direction of the inequality sign.	**Example:** $x + 4 \geq 11$ $x \geq 7$

✍ *Solve each inequality and graph it.*

1) $x + 9 \geq 11$

2) $x - 4 \leq 2$

3) $6x \geq 36$

4) $7 + x < 16$

5) $x + 8 \leq 1$

6) $3x > 12$

7) $3x < 24$

Two–Step Inequalities

Helpful *Hints*	– Isolate the variable. – For dividing both sides by negative numbers, flip the direction of the of the inequality sign. – Simplify using the inverse of addition or subtraction. – Simplify further by using the inverse of multiplication or division.	**Example:** $2x + 9 \geq 11$ $2x \geq 2$ $x \geq 1$

✎ *Solve each inequality and graph it.*

1) $3x - 4 \leq 5$

2) $2x - 2 \leq 6$

3) $4x - 4 \leq 8$

4) $3x + 6 \geq 12$

5) $6x - 5 \geq 19$

6) $2x - 4 \leq 6$

7) $8x - 4 \leq 4$

8) $6x + 4 \leq 10$

9) $5x + 4 \leq 9$

10) $7x - 4 \leq 3$

11) $4x - 19 < 19$

12) $2x - 3 < 21$

13) $7 + 4x \geq 19$

14) $9 + 4x < 21$

15) $3 + 2x \geq 19$

16) $6 + 4x < 22$

Multi–Step Inequalities

Helpful *Hints*	– Isolate the variable. – Simplify using the inverse of addition or subtraction. – Simplify further by using the inverse of multiplication or division.	**Example:** $\dfrac{7x + 1}{3} \geq 5$ $7x + 1 \geq 15$ $7x \geq 14$ $x \geq 7$

✎ *Solve each inequality.*

1) $\dfrac{9x}{7} - 7 < 2$

2) $\dfrac{4x + 8}{2} \leq 12$

3) $\dfrac{3x - 8}{7} > 1$

4) $-3\,(x - 7) > 21$

5) $4 + \dfrac{x}{3} < 7$

6) $\dfrac{2x + 6}{4} \leq 10$

Answers of Worksheets – Chapter 8

Graphing Single–Variable Inequalities

1) $-2 > x$

2) $x \leq -5$

3) $x > 7$

4) $-1.5 > x$

One–Step Inequalities

1)

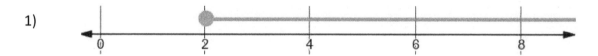

2)

3)

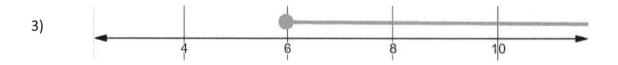

4)
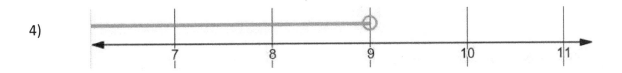

5)

6)
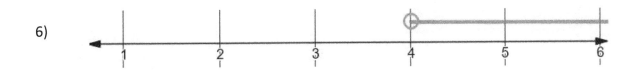

7)

Two–Step inequalities

1) $x \leq 3$

2) $x \leq 4$

3) $x \leq 3$

4) $x \geq 2$

5) $x \geq 4$

6) $x \leq 5$

7) $x \leq 1$

8) $x \leq 1$

9) $x \leq 1$

10) $x \leq 1$

11) $x < 9.5$

12) $x < 12$

13) $x \geq 3$

14) $x < 3$

15) $x \geq 8$

16) $x < 4$

Multi–Step inequalities

1) $x < 7$

2) $x \leq 4$

3) $x > 5$

4) $x < 0$

5) $x < 9$

6) $x \leq 17$

Chapter 9: Linear Functions

Topics that you'll learn in this chapter:

- ✓ Finding Slope
- ✓ Graphing Lines Using Slope– Intercept Form
- ✓ Graphing Lines Using Standard Form
- ✓ Writing Linear Equations
- ✓ Graphing Linear Inequalities
- ✓ Finding Midpoint
- ✓ Finding Distance of Two Points
- ✓ Slope–intercept form
- ✓ Equations of horizontal and vertical lines
- ✓ Equation of parallel or perpendicular lines

"Sometimes the questions are complicated, and the answers are simple." – Dr. Seuss

Finding Slope

Helpful	Slope of a line:	Example:
Hints	$$\frac{y_2 - y_1}{x_2 - x_1} = \frac{rise}{run}$$	$(2, -10), (3, 6)$ slope = 16

✍ *Find the slope of the line through each pair of points.*

1) $(1, 1), (3, 5)$

2) $(4, -6), (-3, -8)$

3) $(7, -12), (5, 10)$

4) $(19, 3), (20, 3)$

5) $(15, 8), (-17, 9)$

6) $(6, -12), (15, -3)$

7) $(3, 1), (7, -5)$

8) $(3, -2), (-7, 8)$

9) $(15, -3), (-9, 5)$

10) $(-4, 7), (-6, -4)$

11) $(6, -8), (-11, -7)$

12) $(-6, 13), (17, -9)$

13) $(-10, -2), (-6, -5)$

14) $(4, 5), (-4, 10)$

15) $(-3, 1), (-17, 2)$

16) $(7, 0), (-13, -11)$

17) $(17, -13), (17, 8)$

18) $(12, 2), (-7, 5)$

Graphing Lines Using Slope–Intercept Form

Helpful	**Slope–intercept form:** given the slope *m* and the y–intercept *b*, then the equation of the line is:
Hints	$y = mx + b.$

Example:

$y = 8x - 3$

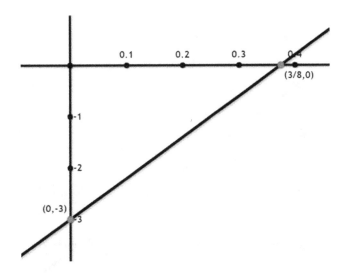

✍ **Sketch the graph of each line.**

1) $y = \dfrac{1}{2} x - 4$

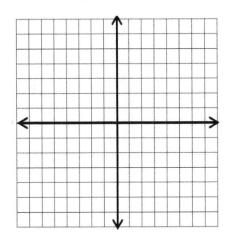

2) $y = 2x$

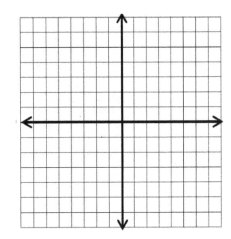

Graphing Lines Using Standard Form

Helpful	– Find the –intercept of the line by putting zero for y.
Hints	– Find the y–intercept of the line by putting zero for the x.
	– Connect these two points.

Example:

$x + 4y = 12$

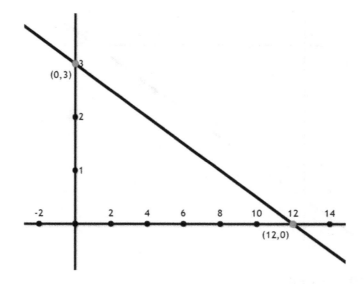

✎ *Sketch the graph of each line.*

1) $2x - y = 4$

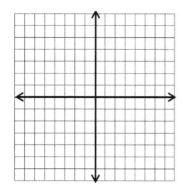

2) $x + y = 2$

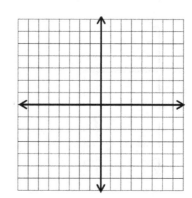

Writing Linear Equations

Helpful	The equation of a line:	**Example:**
	$y = mx + b$	through:
Hints	1– Identify the slope.	$(-4, -2), (-3, 5)$
	2– Find the y–intercept. This can be done by substituting the slope and the coordinates of a point (x, y) on the line.	$y = 7x + 26$

✎ *Write the slope–intercept form of the equation of the line through the given points.*

1) through: $(-4, -2), (-3, 5)$

2) through: $(5, 4), (-4, 3)$

3) through: $(0, -2), (-5, 3)$

4) through: $(-1, 1), (-2, 6)$

5) through: $(0, 3), (-4, -1)$

6) through: $(0, 2), (1, -3)$

7) through: $(0, -5), (4, 3)$

8) through: $(-1, 4), (0, 4)$

9) through: $(2, -3), (3, -5)$

10) through: $(2, 5), (-1, -4)$

11) through: $(1, -3), (-3, 1)$

12) through: $(3, 3), (1, -5)$

13) through: $(4, 4), (3, -5)$

14) through: $(0, 3), (1, 1)$

15) through: $(5, 5), (2, -3)$

16) through: $(-2, -2), (2, -5)$

17) through: $(-3, -2), (1, -1)$

18) through: $(-2, 1), (6, 5)$

Graphing Linear Inequalities

Helpful

Hints

1– First, graph the "equals" line.

2– Choose a testing point. (it can be any point on both sides of the line.)

3– Put the value of (x, y) of that point in the inequality. If that works, that part of the line is the solution. If the values don't work, then the other part of the line is the solution.

✎ *Sketch the graph of each linear inequality.*

1) $y < -4x + 2$

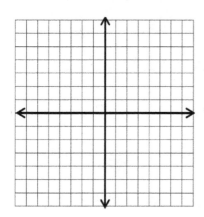

2) $2x + y < -4$

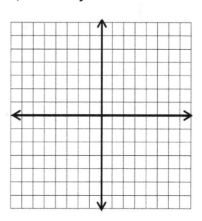

4) $x - 3y < -5$

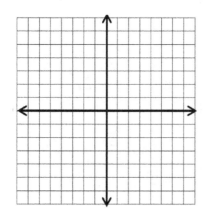

5) $6x - 2y \geq 8$

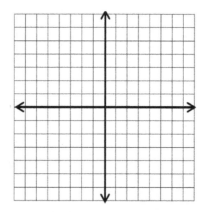

Finding Midpoint

Helpful	Midpoint of the segment AB:	Example:
Hints	$M\left(\dfrac{x_1+x_2}{2}, \dfrac{y_1+y_2}{2}\right)$	$(3, 9), (-1, 6)$ $M\,(1, 7.5)$

✎*Find the midpoint of the line segment with the given endpoints.*

1) $(2, -2), (3, -5)$

2) $(0, 2), (-2, -6)$

3) $(7, 4), (9, -1)$

4) $(4, -5), (0, 8)$

5) $(1, -2), (1, -6)$

6) $(-2, -3), (3, -6)$

7) $(7, 0), (-7, 5)$

8) $(-2, 6), (-3, -2)$

9) $(-1, 1), (5, -5)$

10) $(2.3, -1.3), (-2.2, -0.5)$

11) $(4.1, 6.32), (4, 5.6)$

12) $(2, -1), (-6, 0)$

13) $(-4, 4), (5, -1)$

14) $(-2, -3), (-6, 5)$

15) $\left(\dfrac{1}{2}, 1\right), (2, 4)$

16) $(-2, -2), (6, 5)$

Finding Distance of Two Points

Helpful Hints	Distance from A to B: $$d = \sqrt{(x_1 - x_2)^2 + (y_1 - y_2)^2}$$	Example: $(-1, 2), (-1, -7)$ Distance = 9

✎ **Find the distance between each pair of points.**

1) $(2, -1), (1, -1)$

2) $(6, 4), (-1, 3)$

3) $(-8, -5), (-6, 1)$

4) $(-6, -10), (-2, -10)$

5) $(4, -6), (-3, 4)$

6) $(-6, -7), (-2, -8)$

7) $(5, 4), (8, 2)$

8) $(8, 4), (3, -7)$

9) $(1, 3), (5, 7)$

10) $(4, 2), (-7, 1)$

11) $(-3, -4), (-7, -2)$

12) $(-7, -2), (6, 9)$

13) $(10, 0), (0, 4)$

14) $(-3, 2), (5, 0)$

15) $(-5, 6), (8, -4)$

16) $(3, -5), (-8, -4)$

17) $(0, 8), (4, 10)$

18) $(6, 4), (-5, -1)$

Slope–intercept Form

Helpful *Hints*	Using the slope m and the y-intercept b, then the equation of the line is: $$y = mx + b$$	**Example:** $y = -10 + 2x$ $m = 2$

✍ *Write the slope–intercept form of the equation of each line.*

1) $-14x + y = 7$

2) $-2(2x + y) = 28$

3) $-11x - 7y = -56$

4) $9x + 35 = -5y$

5) $x - 3y = 6$

6) $13x - 11y = -12$

7) $11x - 8y = -48$

8) $3x - 2y = -16$

9) $2y = -6x - 8$

10) $2y = -4x + 10$

11) $2y = -2x - 4$

12) $6x + 5y = -15$

Equations of Horizontal and Vertical Lines

Helpful	The slope of horizontal lines is 0. Thus, the equation of horizontal lines becomes: $y = b$
Hints	The slope of vertical lines is undefined and the equation for a vertical line is: $x = a$

✎ **Sketch the graph of each line.**

1) $y = 0$

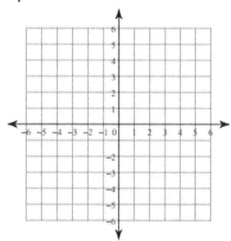

2) $y = 2$

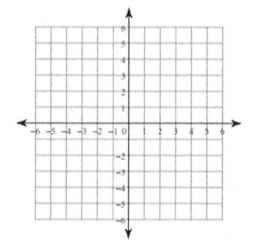

3) $x = -4$

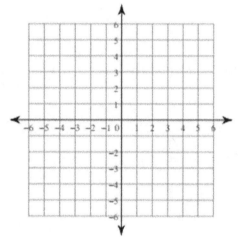

4) $x = 3$

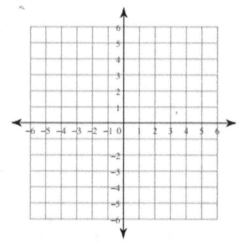

Equation of Parallel or Perpendicular Lines

Helpful *Hints*	**Parallel lines:** are distinct lines with the same slope. For example: if the following lines are parallel: $y = m_1x + b_1$ $y = m_2x + b_2$ Then, $m_1 = m_2$ and $b_1 \neq b_2$. **Perpendicular Lines:** A pair of lines is perpendicular if the lines meet at 90° angle. $y = m_1x + b_1$ $y = m_2x + b_2$ The two lines are perpendicular if, $m_1 = -\frac{1}{m_2}$, that is, if the slopes are negative reciprocals of each other.

✍ *Write an equation of the line that passes through the given point and is parallel to the given line.*

1) $(-2, -4), 4x + 7y = -14$

2) $(-4, 2), y = -x + 3$

3) $(-2, 5), 2y = 4x - 6$

4) $(-10, 0), -y + 3x = 16$

5) $(5, -1), y = -\frac{3}{5}x - 3$

6) $(1, 7), -6x + y = -1$

7) $(2, -3), y = \frac{1}{5}x + 5$

8) $(1, 4), -6x + 5y = -10$

9) $(3, -3), y = -\frac{5}{2}x - 1$

10) $(-4, 3), 2x + 3y = -9$

✍ *Write an equation of the line that passes through the given point and is perpendicular to the given line.*

11) $(-1, -7), 3x + 12y = -6$

12) $(-3, 5), 5x - 6y = 9$

13) $(2, 6), y = -3$

14) $(-2, 3), x = 4$

15) $(1, -5), y = \frac{1}{8}x + 2$

16) $(3, 4), y = -2x - 4$

17) $(-5, 5), y = \frac{5}{9}x - 4$

18) $(4, -1), y = x + 2$

Answers of Worksheets – Chapter 9

Finding Slope

1) 2

2) $\dfrac{2}{7}$

3) −11

4) 0

5) $-\dfrac{1}{32}$

6) 1

7) $-\dfrac{3}{2}$

8) −1

9) $-\dfrac{1}{3}$

10) $\dfrac{11}{2}$

11) $-\dfrac{1}{17}$

12) $-\dfrac{22}{23}$

13) $-\dfrac{3}{4}$

14) $-\dfrac{5}{8}$

15) $-\dfrac{1}{14}$

16) $\dfrac{11}{20}$

17) Undefined

18) $-\dfrac{3}{19}$

Graphing Lines Using Slope–Intercept Form

1)

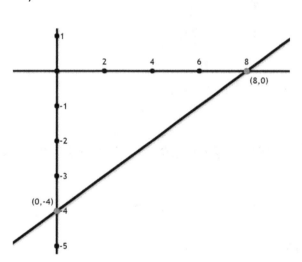

2)

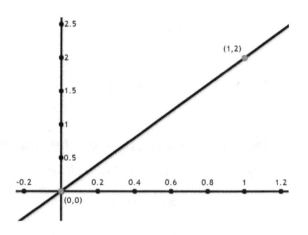

Graphing Lines Using Standard Form

1)

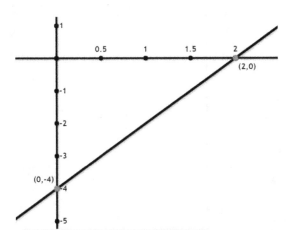

2)

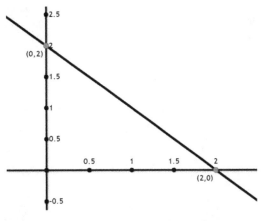

Writing Linear Equations

1) $y = 7x + 26$

2) $y = \frac{1}{9}x + \frac{31}{9}$

3) $y = -x - 2$

4) $y = -5x - 4$

5) $y = x + 3$

6) $y = -5x + 2$

7) $y = 2x - 5$

8) $y = 4$

9) $y = -2x + 1$

10) $y = 3x - 1$

11) $y = -x - 2$

12) $y = 4x - 9$

13) $y = 9x - 32$

14) $y = -2x + 3$

15) $y = \frac{8}{3}x - \frac{25}{3}$

16) $y = -\frac{3}{4}x - \frac{7}{2}$

17) $y = \frac{1}{4}x - \frac{5}{4}$

18) $y = -\frac{4}{3}x + \frac{19}{3}$

Graphing Linear Inequalities

1)

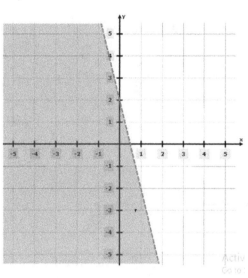

2)

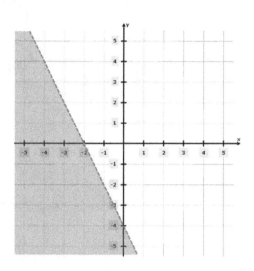

4)

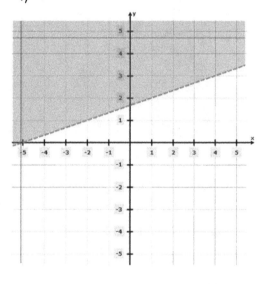

5)

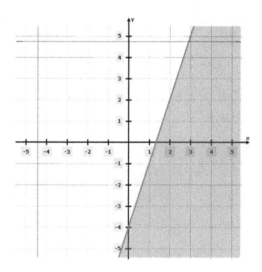

Finding Midpoint

1) (2.5, −3.5)
2) (−1, −2)
3) (8, 1.5)
4) (2, 1.5)
5) (1, −4)
6) (0.5, −4.5)

7) (0, 2.5)
8) (−2.5, 2)
9) (2, −2)
10) (0.05, −0.9)
11) (4.05, 5.96)
12) (−2, − 0.5)

13) $(\frac{1}{2}, 1\frac{1}{2})$
14) (−4, 1)
15) (1.25, 2.5)
16) $(2, \frac{3}{2})$

Finding Distance of Two Points

1) 1
2) 7.1
3) 6.32
4) 4
5) 12.21
6) 4.12

7) 3.61
8) 12.1
9) 5.66
10) 11.04
11) 4.47
12) 17.03

13) 10.77
14) 8.25
15) 16.4
16) 10.3
17) 4.47
18) 12.1

Slope–intercept form

1) $y = 14x + 7$

2) $y = -2x - 14$

3) $y = -\frac{11}{7} x + 8$

4) $y = -\frac{9}{5} x - 7$

5) $y = \frac{x}{3} - 2$

6) $y = \frac{13}{11} x + \frac{12}{11}$

7) $y = \frac{11}{8} x + 6$

8) $y = \frac{3}{2} x + 8$

9) $y = -3x - 4$

10) $y = -2x + 5$

11) $y = - x - 2$

12) $y = -\frac{6}{5} x - 3$

Equations of horizontal and vertical lines

1) $y = 0$ (it is on x axes)

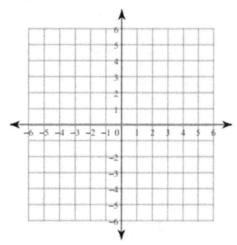

2) $y = 2$

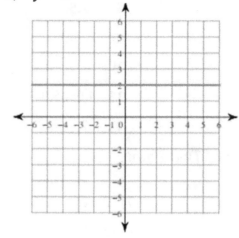

3) $x = -4$

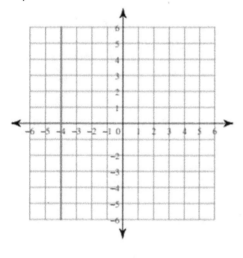

4) $x = 3$

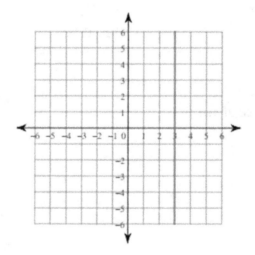

Equation of parallel or perpendicular lines

1) $y = -\dfrac{4}{7}x - \dfrac{36}{7}$

2) $y = -x - 2$

3) $y = 2x + 9$

4) $y = 3x + 30$

5) $y = -\dfrac{3}{5}x + 2$

6) $y = 6x + 1$

7) $y = \dfrac{1}{5}x - \dfrac{17}{5}$

8) $y = \dfrac{6}{5}x + \dfrac{14}{5}$

9) $y = -\dfrac{5}{2}x + \dfrac{9}{2}$

10) $y = -\dfrac{2}{3}x + \dfrac{1}{3}$

11) $y = 4x - 3$

12) $y = -\dfrac{6}{5}x + \dfrac{7}{5}$

13) $x = 2$

14) $y = 3$

15) $y = -8x + 3$

16) $y = \dfrac{1}{2}x + \dfrac{5}{2}$

17) $y = -\dfrac{9}{5}x - 4$

18) $y = -x + 3$

Chapter 10: Polynomials

Topics that you'll learn in this chapter:

✓ Simplifying Polynomials

✓ Multiplying Monomials

✓ Multiplying and Dividing Monomials

✓ Multiplying a Polynomial and a Monomial

✓ Multiplying Binomials

✓ Factoring Trinomials

✓ Operations with Polynomials

✓ Solving Quadratic Equations

Mathematics – the unshaken Foundation of Sciences, and the plentiful Fountain of Advantage to human affairs. — Isaac Barrow

Simplifying Polynomials

Helpful	1– Find "like" terms. (they have same variables with same power).	**Example:**
Hints	2– Add or Subtract "like" terms using PEMDAS operation.	$2x^5 - 3x^3 + 8x^2 - 2x^5 =$ $- 3x^3 + 8x^2$

✍ *Simplify each expression.*

1) $11 - 4x^2 + 3x^2 - 7x^3 + 3$

2) $2x^5 - x^3 + 8x^2 - 2x^5$

3) $(-5)(x^6 + 10) - 8(14 - x^6)$

4) $4(2x^2 + 4x^2 - 3x^3) + 6x^3 + 17$

5) $11 - 6x^2 + 5x^2 - 12x^3 + 22$

6) $2x^2 - 2x + 3x^3 + 12x - 22x$

7) $(3x - 8)(3x - 4)$

8) $(12x + 2y)^2$

9) $(12x^3 + 28x^2 + 10x + 4) \div (x + 2)$

10) $(2x + 12x^2 - 2) \div (2x + 1)$

11) $(2x^3 - 1) + (3x^3 - 2x^3)$

12) $(x - 5)(x - 3)$

13) $(3x + 8)(3x - 8)$

14) $(8x^2 - 3x) - (5x - 5 - 8x^2)$

Multiplying Monomials

Helpful	A monomial is a polynomial with just one term, **Example:**
Hints	like $2x$ or $7y$. $2u^3 \times (-3u)$ $= -6u^4$

✎**Simplify each expression.**

1) $2xy^2z \times 4z^2$

2) $4xy \times x^2y$

3) $4pq^3 \times (-2p^4q)$

4) $8s^4t^2 \times st^5$

5) $12p^3 \times (-3p^4)$

6) $-4p^2q^3r \times 6pq^2r^3$

7) $(-8a^4) \times (-12a^6b)$

8) $3u^4v^2 \times (-7u^2v^3)$

9) $4u^3 \times (-2u)$

10) $-6xy^2 \times 3x^2y$

11) $12y^2z^3 \times (-y^2z)$

12) $5a^2bc^2 \times 2abc^2$

Multiplying and Dividing Monomials

Helpful	- When you divide two monomials you need to divide their coefficients and then divide their variables.
Hints	- In case of exponents with the same base, you need to subtract their powers.

Example:

$(-3x^2)(8x^4y^{12}) = -24x^6y^{12}$

$\dfrac{36\,x^5y^7}{4\,x^4y^5} = 9xy^2$

✎ **Simplify.**

1) $(7x^4y^6)(4x^3y^4)$

2) $(15x^4)\,(3x^9)$

3) $(12x^2y^9)(7x^9y^{12})$

4) $\dfrac{80\,^{12}y^9}{10x^6y^7}$

5) $\dfrac{95\,^{18}y^7}{5x^9y^2}$

6) $\dfrac{200x^3y^8}{40x^3y^7}$

7) $\dfrac{-15x^{17}y^{13}}{3x^6y^9}$

8) $\dfrac{-64x^8y^{10}}{8x^3y^7}$

Multiplying a Polynomial and a Monomial

Helpful	– When multiplying monomials, use the product rule for exponents.	**Example:**
Hints	– When multiplying a monomial by a polynomial, use the distributive property.	$2x(8x - 2) =$
		$16x^2 - 4x$
	$a \times (b + c) = a \times b + a \times c$	

✎**Find each product.**

1) $5(3x - 6y)$

2) $9x(2x + 4y)$

3) $8x(7x - 4)$

4) $12x(3x + 9)$

5) $11x(2x - 11y)$

6) $2x(6x - 6y)$

7) $3x(2x^2 - 3x + 8)$

8) $13x(4x + 8y)$

9) $20(2x^2 - 8x - 5)$

10) $3x(3x - 2)$

11) $6x^3(3x^2 - 2x + 2)$

12) $8x^2(3x^2 - 5xy + 7y^2)$

13) $2x^2(3x^2 - 5x + 12)$

14) $2x^3(2x^2 + 5x - 4)$

15) $5x(6x^2 - 5xy + 2y^2)$

16) $9(x^2 + xy - 8y^2)$

Multiplying Binomials

Helpful	Use "FOIL". (First–Out–In–Last)	**Example:**
Hints	$(x + a)(x + b) = x^2 + (b + a)x + ab$	$(x + 2)(x - 3) =$ $x^2 - x - 6$

✎ *Multiply.*

1) $(3x - 2)(4x + 2)$

2) $(2x - 5)(x + 7)$

3) $(x + 2)(x + 8)$

4) $(x^2 + 2)(x^2 - 2)$

5) $(x - 2)(x + 4)$

6) $(x - 8)(2x + 8)$

7) $(5x - 4)(3x + 3)$

8) $(x - 7)(x - 6)$

9) $(6x + 9)(4x + 9)$

10) $(2x - 6)(5x + 6)$

11) $(x - 7)(x + 7)$

12) $(x + 4)(4x - 8)$

13) $(6x - 4)(6x + 4)$

14) $(x - 7)(x + 2)$

15) $(x - 8)(x + 8)$

16) $(3x + 3)(3x - 4)$

17) $(x + 3)(x + 3)$

18) $(x + 4)(x + 6)$

Factoring Trinomials

Helpful Hints	"FOIL"	Example:
	$(x + a)(x + b) = x^2 + (b + a)x + ab$	$x^2 + 5x + 6 =$
	"Difference of Squares"	$(x + 2)(x + 3)$
	$a^2 - b^2 = (a + b)(a - b)$	
	$a^2 + 2ab + b^2 = (a + b)(a + b)$	
	$a^2 - 2ab + b^2 = (a - b)(a - b)$	
	"Reverse FOIL"	
	$x^2 + (b + a)x + ab = (x + a)(x + b)$	

✎ Factor each trinomial.

1) $x^2 - 7x + 12$

2) $x^2 + 5x - 14$

3) $x^2 - 11x - 42$

4) $6x^2 + x - 12$

5) $x^2 - 17x + 30$

6) $x^2 + 8x + 15$

7) $3x^2 + 11x - 4$

8) $x^2 - 6x - 27$

9) $10x^2 + 33x - 7$

10) $x^2 + 24x + 144$

11) $49x^2 + 28xy + 4y^2$

12) $16x^2 - 40x + 25$

13) $x^2 - 10x + 25$

14) $25x^2 - 20x + 4$

15) $x^3 + 6x^2y^2 + 9xy^3$

16) $9x^2 + 24x + 16$

17) $x^2 - 8x + 16$

18) $x^2 + 121 + 22x$

Operations with Polynomials

Helpful *Hints*	– When multiplying a monomial by a polynomial, use the distributive property. $a \times (b + c) = a \times b + a \times c$	**Example:** $5(6x - 1) =$ $30x - 5$

✍ *Find each product.*

1) $3x^2 (6x - 5)$

2) $5x^2 (7x - 2)$

3) $-3(8x - 3)$

4) $6x^3 (-3x + 4)$

5) $9(6x + 2)$

6) $8(3x + 7)$

7) $5(6x - 1)$

8) $-7x^4 (2x - 4)$

9) $8(x^2 + 2x - 3)$

10) $4(4x^2 - 2x + 1)$

11) $2(3x^2 + 2x - 2)$

12) $8x(5x^2 + 3x + 8)$

13) $(9x + 1)(3x - 1)$

14) $(4x + 5)(6x - 5)$

15) $(7x + 3)(5x - 6)$

16) $(3x - 4)(3x + 8)$

Solving Quadratic Equations

Helpful	Write the equation in the form of $ax^2 + bx + c = 0$
	Factorize the quadratic.
Hints	Use quadratic formula if you couldn't factorize the quadratic.

Quadratic formula

$$x = \frac{-b \pm \sqrt{b^2 - 4ac}}{2a}$$

✍ *Solve each equation by factoring or by using the quadratic formula.*

1) $x^2 + x - 20 = 2x$

2) $x^2 + 8x = -15$

3) $7x^2 - 14x = -7$

4) $6x^2 - 18x - 18 = 6$

5) $2x^2 + 6x - 24 = 12$

6) $2x^2 - 22x + 38 = -10$

7) $(2x + 5)(4x + 3) = 0$

8) $(x + 2)(x - 7) = 0$

9) $(x + 3)(x + 5) = 0$

10) $(5x + 7)(x + 4) = 0$

11) $-4x^2 - 8x - 3 = -3 - 5x^2$

12) $10x^2 = 27x - 18$

13) $7x^2 - 6x + 3 = 3$

14) $x^2 = 2x$

15) $2x^2 - 14 = -3x$

16) $10x^2 - 26x = -12$

17) $15x^2 + 80 = -80x$

18) $x^2 + 15x = -56$

Answers of Worksheets –
Chapter 10

Simplifying Polynomials

1) $-7x^3 - x^2 + 14$

2) $-x^3 + 8x^2$

3) $3x^6 - 162$

4) $-6x^3 + 24x^2 + 17$

5) $-12x^3 - x^2 + 33$

6) $3x^3 + 2x^2 - 12x$

7) $9x^2 - 36x + 32$

8) $144x^2 + 48xy + 4y^2$

9) $12x^2 + 4x + 2$

10) $6x - 1$

11) $3x^3 - 1$

12) $x^2 - 8x + 15$

13) $9x^2 - 64$

14) $16x^2 - 8x + 5$

Multiplying Monomials

1) $8xy^2z^3$

2) $4x^3y^2$

3) $-8p^5q^4$

4) $8s^5t^7$

5) $-36p^7$

6) $-24p^3q^5r^4$

7) $96a^{10}b$

8) $-21u^6v^5$

9) $-8u^4$

10) $-18x^3y^3$

11) $-12y^4z^4$

12) $10a^3b^2c^4$

Multiplying and Dividing Monomials

1) $28x^7y^{10}$

2) $45x^{13}$

3) $84x^{11}y^{21}$

4) $8x^6y^2$

5) $19x^9y^5$

6) $5y$

7) $-5x^{11}y^4$

8) $-8x^5y^3$

Multiplying a Polynomial and a Monomial

1) $15x - 30y$
2) $18x^2 + 36xy$
3) $56x^2 - 32x$
4) $36x^2 + 108x$
5) $22x^2 - 121xy$
6) $12x^2 - 12xy$
7) $6x^3 - 9x^2 + 24x$
8) $52x^2 + 104xy$

9) $40x^2 - 160x - 100$
10) $9x^2 - 6x$
11) $18x^5 - 12x^4 + 12x^3$
12) $24x^4 - 40x^3y + 56y^2x^2$
13) $6x^4 - 10x^3 + 24x^2$
14) $4x^5 + 10x^4 - 8x^3$
15) $30x^3 - 25x^2y + 10xy^2$
16) $9x^2 + 9xy - 72y^2$

Multiplying Binomials

1) $12x^2 - 2x - 4$
2) $2x^2 + 9x - 35$
3) $x^2 + 10x + 16$
4) $x^4 - 4$
5) $x^2 + 2x - 8$
6) $2x^2 - 8x - 64$
7) $15x^2 + 3x - 12$
8) $x^2 - 13x + 42$
9) $24x^2 + 90x + 81$

10) $10x^2 - 18x - 36$
11) $x^2 - 49$
12) $4x^2 + 8x - 32$
13) $36x^2 - 16$
14) $x^2 - 5x - 14$
15) $x^2 - 64$
16) $9x^2 - 3x - 12$
17) $x^2 + 6x + 9$
18) $x^2 + 10x + 24$

Factoring Trinomials

1) $(x - 3)(x - 4)$
2) $(x - 2)(x + 7)$
3) $(x + 3)(x - 14)$
4) $(2x + 3)(3x - 4)$
5) $(x - 15)(x - 2)$
6) $(x + 3)(x + 5)$
7) $(3x - 1)(x + 4)$
8) $(x - 9)(x + 3)$
9) $(5x - 1)(2x + 7)$

10) $(x + 12)(x + 12)$
11) $(7x + 2y)(7x + 2y)$
12) $(4x - 5)(4x - 5)$
13) $(x - 5)(x - 5)$
14) $(5x - 2)(5x - 2)$
15) $x(x^2 + 6xy^2 + 9y^3)$
16) $(3x + 4)(3x + 4)$
17) $(x - 4)(x - 4)$
18) $(x + 11)(x + 11)$

Operations with Polynomials

1) $18x^3 - 15x^2$

2) $35x^3 - 10x^2$

3) $-24x + 9$

4) $-18x^4 + 24x^3$

5) $54x + 18$

6) $24x + 56$

7) $30x - 5$

8) $-14x^5 + 28x^4$

9) $8x^2 + 16x - 24$

10) $16x^2 - 8x + 4$

11) $6x^2 + 4x - 4$

12) $40x^3 + 24x^2 + 64x$

13) $27x^2 - 6x - 1$

14) $24x^2 + 10x - 25$

15) $35x^2 - 27x - 18$

16) $9x^2 + 12x - 32$

Solving Quadratic Equations

1) $\{5, -4\}$

2) $\{-5, -3\}$

3) $\{1\}$

4) $\{4, -1\}$

5) $\{3, -6\}$

6) $\{3, 8\}$

7) $\{-\frac{5}{2}, -\frac{3}{4}\}$

8) $\{-2, 7\}$

9) $\{-3, -5\}$

10) $\{-\frac{7}{5}, -4\}$

11) $\{8, 0\}$

12) $\{\frac{6}{5}, \frac{3}{2}\}$

13) $\{\frac{6}{7}, 0\}$

14) $\{2, 0\}$

15) $\{-\frac{7}{2}, 2\}$

16) $\{\frac{3}{5}, 2\}$

17) $\{-\frac{4}{3}, -4\}$

18) $\{-8, -7\}$

Chapter 11: Exponents and Radicals

Topics that you'll learn in this chapter:

- ✓ Multiplication Property of Exponents
- ✓ Division Property of Exponents
- ✓ Powers of Products and Quotients
- ✓ Zero and Negative Exponents
- ✓ Negative Exponents and Negative Bases
- ✓ Writing Scientific Notation
- ✓ Square Roots

Mathematics is no more computation than typing is literature.

— John Allen Paulos

Multiplication Property of Exponents

Helpful	Exponents rules	Example:
Hints	$x^a \cdot x^b = x^{a+b}$ $\dfrac{x^a}{x^b} = x^{a-b}$ $\dfrac{1}{x^b} = x^{-b}$ $(x^a)^b = x^{a.b}$ $(xy)^a = x^a \cdot y^a$	$(x^2 y)^3 = x^6 y^3$

✎ *Simplify.*

1) $4^2 \cdot 4^2$

2) $2 \cdot 2^2 \cdot 2^2$

3) $3^2 \cdot 3^2$

4) $3x^3 \cdot x$

5) $12x^4 \cdot 3x$

6) $6x \cdot 2x^2$

7) $5x^4 \cdot 5x^4$

8) $6x^2 \cdot 6x^3 y^4$

9) $7x^2 y^5 \cdot 9xy^3$

10) $7xy^4 \cdot 4x^3 y^3$

11) $(2x^2)^2$

12) $3x^5 y^3 \cdot 8x^2 y^3$

13) $7x^3 \cdot 10y^3 x^5 \cdot 8yx^3$

14) $(x^4)^3$

15) $(2x^2)^4$

16) $(x^2)^3$

17) $(6x)^2$

18) $3x^4 y^5 \cdot 7x^2 y^3$

Division Property of Exponents

Helpful	$\dfrac{x^a}{x^b} = x^{a-b}$, $x \neq 0$	Example:
Hints		$\dfrac{x^{12}}{x^5} = x^7$

✎ *Simplify.*

1) $\dfrac{5^5}{5}$

2) $\dfrac{3}{3^5}$

3) $\dfrac{2^2}{2^3}$

4) $\dfrac{2^4}{2^2}$

5) $\dfrac{x}{x^3}$

6) $\dfrac{3x^3}{9x^4}$

7) $\dfrac{2x^{-5}}{9x^{-2}}$

8) $\dfrac{21x^8}{7x^3}$

9) $\dfrac{7x^6}{4x^7}$

10) $\dfrac{6x^2}{4x^3}$

11) $\dfrac{5x}{10x^3}$

12) $\dfrac{3x^3}{2x^5}$

13) $\dfrac{12x^3}{14x^6}$

14) $\dfrac{12x^3}{9y^8}$

15) $\dfrac{25xy^4}{5x^6y^2}$

16) $\dfrac{2x^4}{7x}$

17) $\dfrac{16\ ^2y^8}{4x^3}$

18) $\dfrac{12\ ^4}{15x^7y^9}$

19) $\dfrac{12yx^4}{10yx^8}$

20) $\dfrac{16\ ^4y}{9x^8y^2}$

21) $\dfrac{5x^8}{20x^8}$

Powers of Products and Quotients

Helpful	For any nonzero numbers a and b and any integer $(ab)^x = a^x b^x$	**Example:**
Hints		$(2x^2 . y^3)^2 =$ $4x^2 . y^6$

✍ *Simplify.*

1) $(2x^3)^4$

2) $(4xy^4)^2$

3) $(5x^4)^2$

4) $(11x^5)^2$

5) $(4x^2y^4)^4$

6) $(2x^4y^4)^3$

7) $(3x^2y^2)^2$

8) $(3x^4y^3)^4$

9) $(2x^6y^8)^2$

10) $(12x\ 3x)^3$

11) $(2x^9\ x^6)^3$

12) $(5x^{10}y^3)^3$

13) $(4x^3\ x^2)^2$

14) $(3x^3\ 5x)^2$

15) $(10x^{11}y^3)^2$

16) $(9x^7\ y^5)^2$

17) $(4x^4y^6)^5$

18) $(4x^4)^2$

19) $(3x\ 4y^3)^2$

20) $(9x^2y)^3$

21) $(12x^2y^5)^2$

Zero and Negative Exponents

Helpful Hints	A negative exponent simply means that the base is on the wrong side of the fraction line, so you need to flip the base to the other side. For instance, "x^{-2}" (pronounced as "ecks to the minus two") just means "x^2" but underneath, as in $\frac{1}{x^2}$	**Example:** $5^{-2} = \frac{1}{25}$

🖎 *Evaluate the following expressions.*

1) 8^{-2}

2) 2^{-4}

3) 10^{-2}

4) 5^{-3}

5) 22^{-1}

6) 9^{-1}

7) 3^{-2}

8) 4^{-2}

9) 5^{-2}

10) 35^{-1}

11) 6^{-3}

12) 0^{15}

13) 10^{-9}

14) 3^{-4}

15) 5^{-2}

16) 2^{-3}

17) 3^{-3}

18) 8^{-1}

19) 7^{-3}

20) 6^{-2}

21) $\left(\frac{2}{3}\right)^{-2}$

22) $\left(\frac{1}{5}\right)^{-3}$

23) $\left(\frac{1}{2}\right)^{-8}$

24) $\left(\frac{2}{5}\right)^{-3}$

25) 10^{-3}

26) 1^{-10}

Negative Exponents and Negative Bases

Helpful	– Make the power positive. A negative exponent is the reciprocal of that number with a positive exponent.	**Example:**
Hints	– The parenthesis is important!	$2x^{-3} = \dfrac{2}{x^3}$
	-5^{-2} is not the same as $(-5)^{-2}$	
	$-5^{-2} = -\dfrac{1}{5^2}$ and $(-5)^{-2} = +\dfrac{1}{5^2}$	

✎ *Simplify.*

1) -6^{-1}

2) $-4x^{-3}$

3) $-\dfrac{5x}{x^{-3}}$

4) $-\dfrac{a^{-3}}{b^{-2}}$

5) $-\dfrac{5}{x^{-3}}$

6) $\dfrac{7b}{-9c^{-4}}$

7) $-\dfrac{5n^{-2}}{10p^{-3}}$

8) $\dfrac{4ab^{-2}}{-3c^{-2}}$

9) $-12x^2y^{-3}$

10) $\left(-\dfrac{1}{3}\right)^{-2}$

11) $\left(-\dfrac{3}{4}\right)^{-2}$

12) $\left(\dfrac{3a}{2c}\right)^{-2}$

13) $\left(-\dfrac{5x}{3y}\right)^{-3}$

14) $-\dfrac{2x}{a^{-4}}$

Writing Scientific Notation

Helpful	– It is used to write very big or very small numbers in decimal form.

Hints

– In scientific notation all numbers are written in the form of:

$$m \times 10^n$$

Decimal notation	Scientific notation
5	5×10^0
−25,000	-2.5×10^4
0.5	5×10^{-1}
2,122.456	$2,122456 \times 10^3$

✍ *Write each number in scientific notation.*

1) 91×10^3

2) 60

3) 2000000

4) 0.0000006

5) 354000

6) 0.000325

7) 2.5

8) 0.00023

9) 56000000

10) 2000000

11) 78000000

12) 0.0000022

13) 0.00012

14) 0.004

15) 78

16) 1600

17) 1450

18) 130000

19) 60

20) 0.113

21) 0.02

Square Roots

Helpful	− A square root of x is a number r whose square is: $r^2 = x$	Example:
Hints	r is a square root of x.	$\sqrt{4} = 2$

✍️*Find the value each square root.*

1) $\sqrt{1}$

2) $\sqrt{4}$

3) $\sqrt{9}$

4) $\sqrt{25}$

5) $\sqrt{16}$

6) $\sqrt{49}$

7) $\sqrt{36}$

8) $\sqrt{0}$

9) $\sqrt{64}$

10) $\sqrt{81}$

11) $\sqrt{121}$

12) $\sqrt{225}$

13) $\sqrt{144}$

14) $\sqrt{100}$

15) $\sqrt{256}$

16) $\sqrt{289}$

17) $\sqrt{324}$

18) $\sqrt{400}$

19) $\sqrt{900}$

20) $\sqrt{529}$

21) $\sqrt{90}$

Answers of Worksheets – Chapter 11

Multiplication Property of Exponents

1) 4^4

2) 2^5

3) 3^4

4) $3x^4$

5) $36x^5$

6) $12x^3$

7) $25x^8$

8) $36x^5y^4$

9) $63x^3y^8$

10) $28x^4y^7$

11) $4x^4$

12) $24x^7y^6$

13) $560x^{11}y^4$

14) x^{12}

15) $16x^8$

16) x^6

17) $36x^2$

18) $21x^6y^8$

Division Property of Exponents

1) 5^4

2) $\frac{1}{3^4}$

3) $\frac{1}{2}$

4) 2^2

5) $\frac{1}{x^2}$

6) $\frac{1}{3x}$

7) $\frac{2}{9x^3}$

8) $3x^5$

9) $\frac{7}{4x}$

10) $\frac{3}{2x}$

11) $\frac{1}{2x^2}$

12) $\frac{3}{2x^2}$

13) $\frac{6}{7x^3}$

14) $\frac{4x^3}{3y^8}$

15) $\frac{5y^2}{x^5}$

16) $\frac{2x^3}{7}$

17) $\frac{4y^8}{x}$

18) $\frac{4}{5x^3y^9}$

19) $\frac{6}{5x^4}$

20) $\frac{16}{9x^4y}$

21) $\frac{1}{4}$

Powers of Products and Quotients

1) $16x^{12}$

2) $16x^2y^8$

3) $25x^8$

4) $121x^{10}$

5) $256x^8y^{16}$

6) $8x^{12}y^{12}$

7) $9x^4y^4$

8) $81x^{16}y^{12}$

9) $4x^{12}y^{16}$

10) $46,656x^6$

11) $8x^{45}$

12) $125x^{30}y^9$

13) $16x^{10}$

14) $225x^8$

15) $100x^{22}y^6$

16) $81x^{14}y^{10}$

17) $1,024x^{20}y^{30}$

18) $16x^8$

19) $144x^2y^6$ 20) $729x^6y^3$ 21) $144x^4y^{10}$

Zero and Negative Exponents

1) $\frac{1}{64}$

2) $\frac{1}{16}$

3) $\frac{1}{100}$

4) $\frac{1}{125}$

5) $\frac{1}{22}$

6) $\frac{1}{9}$

7) $\frac{1}{9}$

8) $\frac{1}{16}$

9) $\frac{1}{25}$

10) $\frac{1}{35}$

11) $\frac{1}{216}$

12) 0

13) $\frac{1}{1000000000}$

14) $\frac{1}{81}$

15) $\frac{1}{25}$

16) $\frac{1}{8}$

17) $\frac{1}{27}$

18) $\frac{1}{8}$

19) $\frac{1}{343}$

20) $\frac{1}{36}$

21) $\frac{9}{4}$

22) 125

23) 256

24) $\frac{125}{8}$

25) $\frac{1}{1000}$

26) 1

Negative Exponents and Negative Bases

1) $-\frac{1}{6}$

2) $-\frac{4}{x^3}$

3) $-5x^4$

4) $-\frac{b^2}{a^3}$

5) $-5x^3$

6) $-\frac{7bc^4}{9}$

7) $-\frac{p^3}{2n^2}$

8) $-\frac{4ac^2}{3b^2}$

9) $-\frac{12^2}{y^3}$

10) 9

11) $\frac{16}{9}$

12) $\frac{4c^2}{9a^2}$

13) $-\frac{27y^3z^3}{125x^3}$

14) $-2xa^4$

Writing Scientific Notation

1) 9.1×10^4

2) 6×10^1

3) 2×10^6

4) 6×10^{-7}

5) 3.54×10^5

6) 3.25×10^{-4}

7) 2.5×10^0

8) 2.3×10^{-4}

9) 5.6×10^7

10) 2×10^6

11) 7.8×10^7

12) 2.2×10^{-6}

13) 1.2×10^{-4} 16) 1.6×10^{3} 19) 6×10^{1}

14) 4×10^{-3} 17) 1.45×10^{3} 20) 1.13×10^{-1}

15) 7.8×10^{1} 18) 1.3×10^{5} 21) 2×10^{-2}

Square Roots

1) 1   8) 0   15) 16

2) 2   9) 8   16) 17

3) 3   10) 9   17) 18

4) 5   11) 11   18) 20

5) 4   12) 15   19) 30

6) 7   13) 12   20) 23

7) 6   14) 10   21) $3\sqrt{10}$

Chapter 12: Geometry

Topics that you'll learn in this chapter:

✓ The Pythagorean Theorem

✓ Area of Triangles

✓ Perimeter of Polygons

✓ Area and Circumference of Circles

✓ Area of Squares, Rectangles, and Parallelograms

✓ Area of Trapezoids

Mathematics is, as it were, a sensuous logic, and relates to philosophy as do the arts, music, and plastic art to poetry. — K. Shegel

The Pythagorean Theorem

Helpful *Hints*	– In any right triangle: $a^2 + b^2 = c^2$	**Example:** Missing side = 6 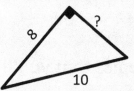

✎ **Do the following lengths form a right triangle?**

1)

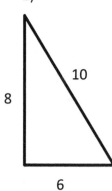

8, 10, 6

2)

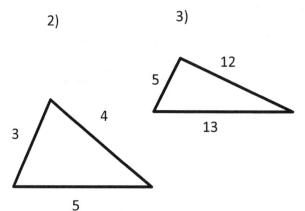

3, 4, 5

3)

5, 12, 13

✎ **Find each missing length to the nearest tenth.**

4)

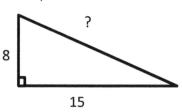

8, ?, 15

5)

24, ?, 10

6)

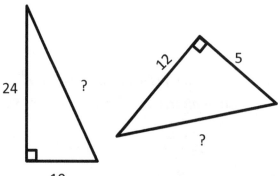

12, 5, ?

Area of Triangles

Helpful

Hints

Area $= \dfrac{1}{2}$ $(base \times height)$

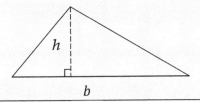

✏️*Find the area of each.*

1)

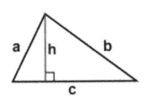

c = 9 mi

h = 3.7 mi

2)

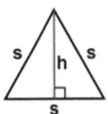

s = 14 m

h = 12.2 m

3)

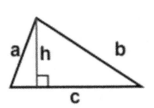

a = 5 m

b = 11 m

c = 14 m

h = 4 m

4)

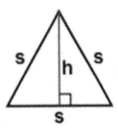

s = 10 m

h = 8.6 m

Perimeter of Polygons

Helpful

Hints

Perimeter of a square = 4s

  s

Perimeter of a rectangle

= $2(l + w)$

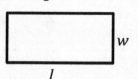 w

l

Perimeter of trapezoid

= a + b + c + d a

  b

d

c

Perimeter of Pentagon = 6a

  a

Perimeter of a parallelogram = $2(l + w)$

l

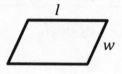 w

Example:

P = 18

3 m

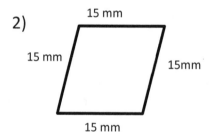

3 m   3 m

✎ **Find the perimeter of each shape.**

1)

5 m

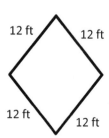

5 m   5 m

2)

15 mm

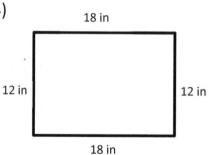

15 mm   15mm

15 mm

3)

12 ft   12 ft

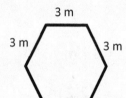

12 ft   12 ft

4)

18 in

12 in   12 in

18 in

Area and Circumference of Circles

Area = πr²

Circumference = 2πr

Example:

If the radius of a circle is 3, then:

Area = 28.27

Circumference = 18.85

✎ **Find the area and circumference of each.** (π = 3.14)

1)

2)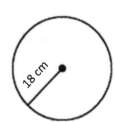

3)

4)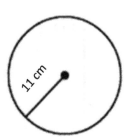

5)

6)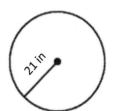

Area of Squares, Rectangles, and Parallelograms

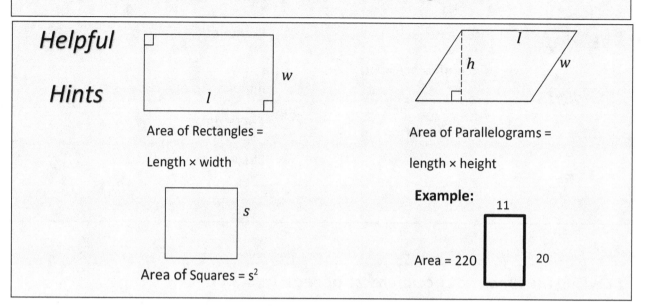

Helpful

Hints

Area of Rectangles =

Length × width

Area of Squares = s²

Area of Parallelograms =

length × height

Example:

11

Area = 220   20

✎ *Find the area of each.*

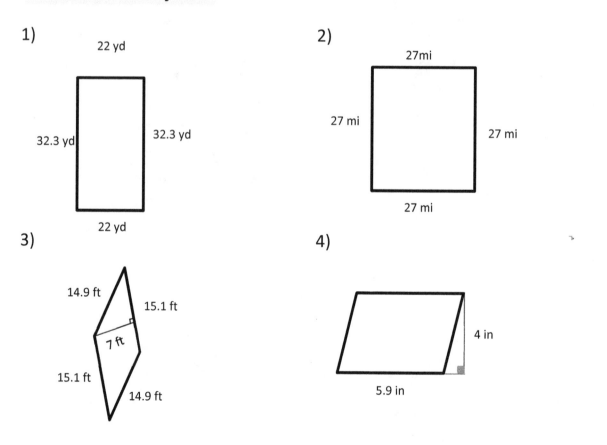

1)

22 yd

32.3 yd   32.3 yd

22 yd

2)

27mi

27 mi   27 mi

27 mi

3)

14.9 ft

15.1 ft

7 ft

15.1 ft

14.9 ft

4)

4 in

5.9 in

Area of Trapezoids

Helpful

$$A = \frac{1}{2}h(b_1 + b_2)$$

Hints

Example:

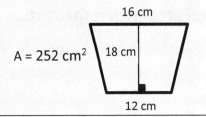

A = 252 cm²

📝 *Calculate the area for each trapezoid.*

1)

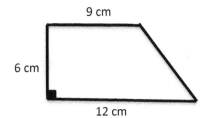

2)

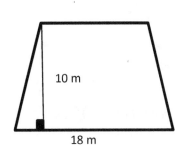

3)

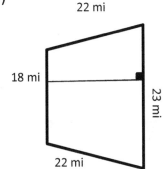

4)

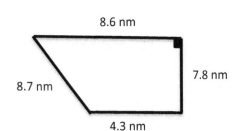

Answers of Worksheets – Chapter 12

The Pythagorean Theorem

1) yes

2) yes

3) yes

4) 17

5) 26

6) 13

Area of Triangles

1) 16.65 mi^2

2) 85.4 m^2

3) 28 m^2

4) 43 m^2

Perimeter of Polygons

1) 30 m

2) 60 mm

3) 48 ft

4) 60 in

Area and Circumference of Circles

1) Area: 50.24 in^2, Circumference: 25.12 in

2) Area: 1,017.36 cm^2, Circumference: 113.04 cm

3) Area: 78.5m^2, Circumference: 31.4 m

4) Area: 379.94 cm^2, Circumference: 69.08 cm

5) Area: 200.96 km^2, Circumference: 50.2 km

6) Area: 1,384.74 km^2, Circumference: 131.88 km

Area of Squares, Rectangles, and Parallelograms

1) 710.6 yd^2

2) 729 mi^2

3) 105.7 ft^2

4) 23.6 in^2

Area of Trapezoids

1) 63 cm^2

2) 160 m^2

3) 410 mi^2

4) 50.31 nm^2

Chapter 13: Solid Figures

Topics that you'll learn in this chapter:

- ✓ Volume of Cubes
- ✓ Volume of Rectangle Prisms
- ✓ Surface Area of Cubes
- ✓ Surface Area of Rectangle Prisms
- ✓ Volume of a Cylinder
- ✓ Surface Area of a Cylinder

Mathematics is a great motivator for all humans. Because its career starts with zero and it never end (infinity)

Volume of Cubes

Helpful	– Volume is the measure of the amount of space inside of a solid figure, like a cube, ball, cylinder or pyramid.
Hints	– Volume of a cube = (one side)^3
	– Volume of a rectangle prism: Length × Width × Height

✎ *Find the volume of each.*

1)

2)

3)

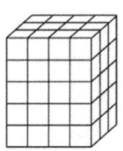

4)

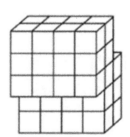

5)

6)

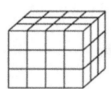

Volume of Rectangle Prisms

Helpful		Example:
	Volume of rectangle prism	$10 \times 5 \times 8 = 400m^3$
Hints	length × width × height	

✎ *Find the volume of each of the rectangular prisms.*

1)

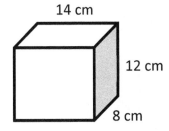

2)

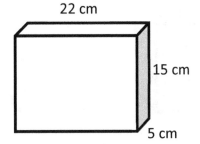

3)

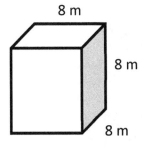

4)

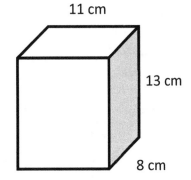

Surface Area of Cubes

Helpful

Hints

Surface Area of a cube =

6 × (one side of the cube)^2

Example:

$6 × 4^2 = 96m^2$

4 m

4 m

4 m

✎ *Find the surface of each cube.*

1)

6 mm

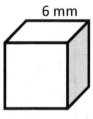

2)

9 mm

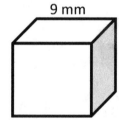

3)

10 cm

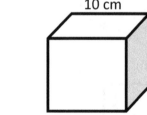

4)

8 m

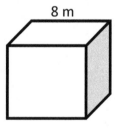

5)

7.5 in

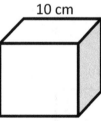

6)

11.3 ft

Surface Area of a Rectangle Prism

Helpful

Hints

Surface Area of a Rectangle Prism Formula:

SA =2 [(width × length) + (height × length) + width × height)]

✎ *Find the surface of each prism.*

1)

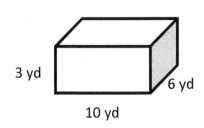

3 yd
6 yd
10 yd

2)

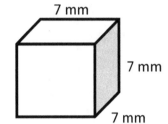

7 mm
7 mm
7 mm

3)

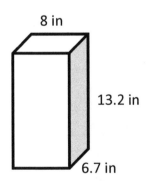

8 in
13.2 in
6.7 in

4)

17 cm

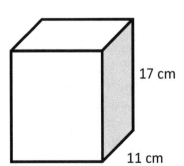

17 cm
11 cm

Volume of a Cylinder

> **Helpful**
>
> **Hints**
>
> Volume of Cylinder Formula = π(radius)^2 × height
>
> π = 3.14

✎ *Find the volume of each cylinder.* (π = 3.14)

1)

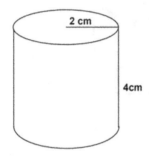

2 cm

4cm

2)

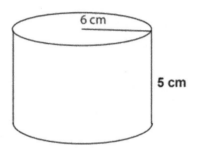

6 cm

5 cm

3)

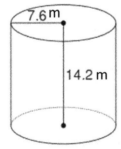

7.6 m

14.2 m

4)

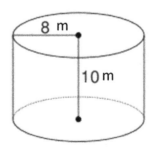

8 m

10 m

Surface Area of a Cylinder

Helpful	Surface area of a cylinder	Example:
Hints	$SA = 2\pi r^2 + 2\pi rh$	Surface area = 1727

14 m

11 m

✏️ **Find the surface of each cylinder.** ($\pi = 3.14$)

1)

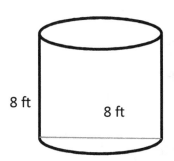

8 ft

8 ft

2)

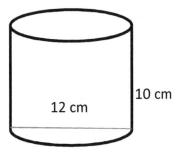

10 cm

12 cm

3)

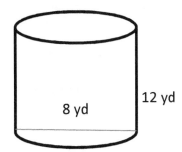

16 in

18 in

4)

12 yd

8 yd

Answers of Worksheets – Chapter 13

Volumes of Cubes

1) 8
2) 4
3) 5
4) 36
5) 60
6) 44

Volume of Rectangle Prisms

1) 1344 cm^3
2) 1650 cm^3
3) 512 m^3
4) 1144 cm^3

Surface Area of a Cube

1) 216 mm^2
2) 486 mm^2
3) 600 cm^2
4) 384 m^2
5) 337.5 in^2
6) 766.14 ft^2

Surface Area of a Prism

1) 216 yd^2
2) 294 mm^2
3) 495.28 in^2
4) 1326 cm^2

Volume of a Cylinder

1) 50.24 cm^3
2) 565.2 cm^3
3) 2,575.403 m^3
4) 2009.6 m^3

Surface Area of a Cylinder

1) 301.44 ft^2
2) 602.88 cm^2
3) 1413 in^2
4) 401.92 yd^2

Chapter 14: Statistics

Topics that you'll learn in this chapter:

- ✓ Mean, Median, Mode, and Range of the Given Data
- ✓ Box and Whisker Plots
- ✓ Bar Graph
- ✓ Stem– And– Leaf Plot
- ✓ The Pie Graph or Circle Graph
- ✓ Scatter Plots
- ✓ Probability

Mathematics is no more computation than typing is literature.

— John Allen Paulos

Mean, Median, Mode, and Range of the Given Data

Helpful		Example:
	- Mean: $\dfrac{\text{sum of the data}}{\text{of data entires}}$	
Hints	- Mode: value in the list that appears most often	22, 16, 12, 9, 7, 6, 4, 6
	- Range: largest value – smallest value	Mean = 10.25
		Mod = 6
		Range = 18

✎ *Find Mean, Median, Mode, and Range of the Given Data.*

1) 7, 2, 5, 1, 1, 2

2) 2, 2, 2, 3, 6, 3, 7, 4

3) 9, 4, 3, 1, 7, 9, 4, 6, 4

4) 8, 4, 2, 4, 3, 2, 4, 5

5) 8, 5, 7, 5, 7, 9, 8

6) 5, 1, 4, 4, 9, 2, 9, 2, 5, 1

7) 4, 1, 5, 9, 7, 7, 5, 4, 3, 5

8) 7, 5, 4, 9, 6, 7, 7, 5, 2

9) 2, 5, 5, 6, 2, 4, 7, 6, 4, 9

10) 10, 5, 2, 5, 4, 5, 8, 10

11) 5, 1, 5, 2, 2

12) 2, 3, 5, 9, 6

Bar Graph

Helpful *Hints*	— A bar graph is a chart that presents data with bars in different heights to match with the values of the data. The bars can be graphed horizontally or vertically.

✎*Graph the given information as a bar graph.*

Day	Hot dogs sold
Monday	90
Tuesday	70
Wednesday	30
Thursday	20
Friday	60

The Pie Graph or Circle Graph

Helpful	A Pie Chart is a circle chart divided into sectors, each sector represents the relative size of each value.
Hints	

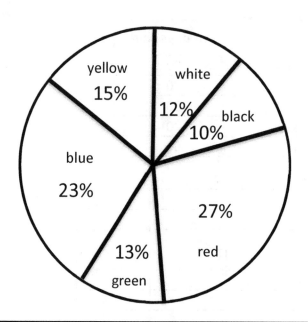

Favorite colors

1) Which color is the most popular?

2) What percentage of pie graph is yellow?

3) Which color is the least popular?

4) What percentage of pie graph is blue?

5) What percentage of pie graph is green?

Scatter Plots

Helpful	A Scatter (xy) Plot shows the values with points that represent the relationship between two sets of data.
Hints	– The horizontal values are usually x and vertical data is y.

✎ *Construct a scatter plot.*

X	Y
1	20
2	40
3	50
4	60

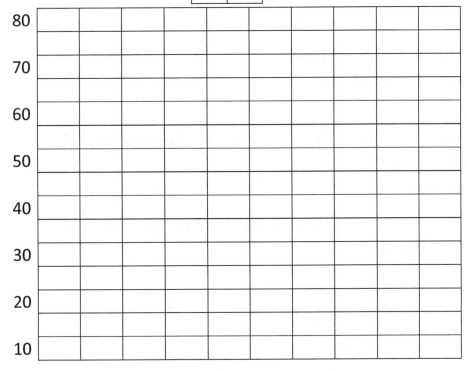

Probability Problems

Helpful Hints	- Probability is the likelihood of something happening in the future. It is expressed as a number between zero (can never happen) to 1 (will always happen). - Probability can be expressed as a fraction, a decimal, or a percent.	**Example:** Probability of a flipped coins turns up 'heads' Is $0.5 = \frac{1}{2}$

✎ *Solve.*

1) A number is chosen at random from 1 to 10. Find the probability of selecting a 4 or smaller.

2) A number is chosen at random from 1 to 50. Find the probability of selecting multiples of 10.

3) A number is chosen at random from 1 to 10. Find the probability of selecting of 4 and factors of 6.

4) A number is chosen at random from 1 to 10. Find the probability of selecting a multiple of 3.

5) A number is chosen at random from 1 to 50. Find the probability of selecting prime numbers.

6) A number is chosen at random from 1 to 25. Find the probability of not selecting a composite number.

Factorials

Helpful Hints	Factorials means to multiply a series of descending natural numbers.	**Example:** $4! = 4 \times 3 \times 2 \times 1$

✎ *Determine the value for each expression.*

1) $\dfrac{9!}{6!}$

2) $\dfrac{8!}{5!}$

3) $\dfrac{7!}{5!}$

4) $\dfrac{20!}{18!}$

5) $\dfrac{22!}{18!5!}$

6) $\dfrac{10!}{8!2!}$

7) $\dfrac{100!}{97!}$

8) $\dfrac{14!}{10!4!}$

9) $\dfrac{10!}{8!}$

10) $\dfrac{25!}{20!}$

11) $\dfrac{14!}{9!3!}$

12) $\dfrac{55!}{53!}$

13) $\dfrac{(2.3)!}{3!}$

14) $5! + 4!$

Answers of Worksheets – Chapter 14

Mean, Median, Mode, and Range of the Given Data

1) mean: 3, median: 2, mode: 1, 2, range: 6
2) mean: 3.625, median: 3, mode: 2, range: 5
3) mean: 5.22, median: 4, mode: 4, range: 8
4) mean: 4, median: 4, mode: 4, range: 6
5) mean: 7, median: 7, mode: 5, 7, 8, range: 4
6) mean: 4.2, median: 4, mode: 1,2,4,5,9, range: 8
7) mean: 5, median: 5, mode: 5, range: 8
8) mean: 5.78, median: 6, mode: 7, range: 7
9) mean: 5, median: 5, mode: 2, 4, 5, 6, range: 7
10) mean: 6.125, median: 5, mode: 5, range: 8
11) mean: 3, median: 2, mode: 2, 5, range: 4
12) mean: 5, median: 5, mode: none, range: 7

Bar Graph

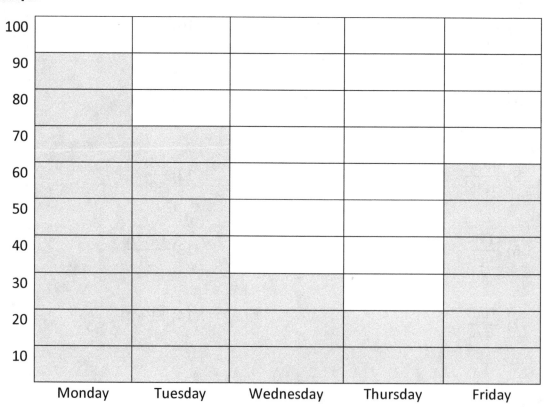

The Pie Graph or Circle Graph

1) red

2) 15%

3) black

4) 23%

5) 13%

Scatter Plots

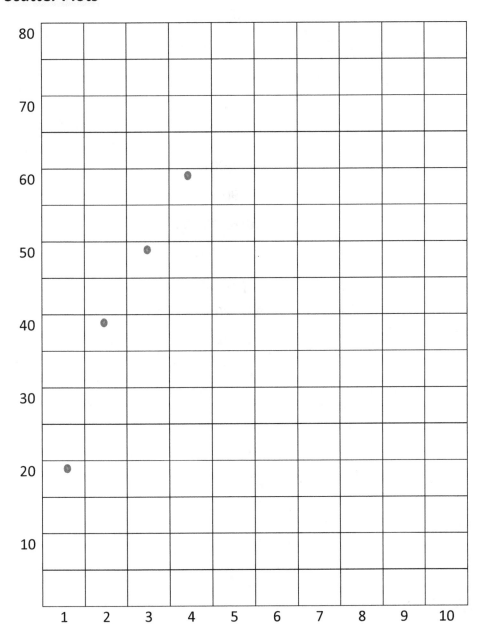

Probability Problems

1) $\dfrac{2}{5}$

2) $\dfrac{1}{10}$

3) $\dfrac{1}{2}$

4) $\dfrac{3}{10}$

5) $\dfrac{7}{25}$

6) $\dfrac{9}{25}$

Factorials

1) 504

2) 336

3) 42

4) 380

5) 1,463

6) 45

7) 970,200

8) 1,001

9) 90

10) 6,375,600

11) 40,040

12) 2,970

13) 120

14) 144

AFOQT Math Practice Tests

The Air Force Officer Qualifying Test (AFOQT) is a standardized test to assess skills and personality traits that have proven to be predictive of success in officer commissioning programs such as the training program.

The AFOQT is used to select applicants for officer commissioning programs, such as Officer Training School (OTS) or Air Force Reserve Officer Training Corps (Air Force ROTC) and pilot and navigator training.

The AFOQT is a multiple-aptitude battery that measures developed abilities and helps predict future academic and occupational success in the military. The AFOQT is a multiple-choice test which consists of 12 subtests and two of them are Arithmetic Reasoning and Mathematics Knowledge.

In this section, there are 10 complete Arithmetic Reasoning and Mathematics Knowledge AFOQT Tests. Take these tests to see what score you'll be able to receive on a real AFOQT test.

Good luck!

Test 1

Arithmetic Reasoning

- o **25 questions**
- o **Total time for this section:** 29 Minutes
- o **Calculators are not allowed at the test.**

1) Aria was hired to teach three identical math courses, which entailed being present in the classroom 36 hours altogether. At $25 per class hour, how much did Aria earn for teaching one course?

A. $50

C. $600

B. $300

D. $1,400

2) Karen is 9 years older than her sister Michelle, and Michelle is 4 years younger than her brother David. If the sum of their ages is 82, how old is Michelle?

A. 21

C. 29

B. 25

D. 23

3) John is driving to visit his mother, who lives 300 miles away. How long will the drive be, round–trip, if John drives at an average speed of 50 mph?

A. 95 Minutes

C. 645 Minutes

B. 260 Minutes

D. 720 Minutes

4) Julie gives 8 pieces of candy to each of her friends. If Julie gives all her candy away, which amount of candy could have been the amount she distributed?

A. 187

C. 343

B. 216

D. 223

5) If a rectangle is 30 feet by 45 feet, what is its area?

A. 1,350

C. 1,000

B. 1,250

D. 750

6) You are asked to chart the temperature during an 8 hour period to give the average. These are your results:

7 am: 2 degrees   11 am: 32 degrees

8 am: 5 degrees   12 pm: 35 degrees

9 am: 22 degrees   1 pm: 35 degrees

10 am: 28 degrees   2 pm: 33 degrees

What is the average temperature?

A. 36   C. 24

B. 28   D. 22

7) Each year, a cyber café charges its customers a base rate of $15, with an additional $0.20 per visit for the first 40 visits, and $0.10 for every visit after that. How much does the cyber café charge a customer for a year in which 60 visits are made?

A. $25   C. $35

B. $29   D. $39

8) If a vehicle is driven 32 miles on Monday, 35 miles on Tuesday, and 29 miles on Wednesday, what is the average number of miles driven each day?

A. 32 Miles   C. 29 Miles

B. 31 Miles   D. 27 Miles

9) Three co-workers contributed $10.25, $11.25, and $18.45 respectively to purchase a retirement gift for their boss. What is the maximum amount they can spend on a gift?

A. $42.25   C. $39.95

B. $40.17   D. $27.06

10) While at work, Emma checks her email once every 90 minutes. In 9−hour, how many times does she check her email?

    A. 4 Times   C. 7 Times

    B. 5 Times   D. 6 Times

11) A family owns 15 dozen of magazines. After donating 57 magazines to the public library, how many magazines are still with the family?

    A. 180   C. 123

    B. 152   D. 98

12) In the deck of cards, there are 4 spades, 3 hearts, 7 clubs, and 10 diamonds. What is the probability that William will pick out a spade?

    A. 1/6   C. 1/9

    B. 1/8   D. 1/5

13) What is the prime factorization of 560?

    A. $2 \times 2 \times 5 \times 7$ C. 2×7

    B. $2 \times 2 \times 2 \times 2 \times 5 \times 7$ D. $2 \times 2 \times 2 \times 5 \times 7$

14) William is driving a truck that can hold 5 tons maximum. He has a shipment of food weighing 32,000 pounds. How many trips will he need to make to deliver all of the food?

    A. 1 Trip   C. 4 Trips

    B. 3 Trips   D. 6 Trips

15) A man goes to a casino with $180. He loses $40 on blackjack, then loses another $50 on roulette. How much money does he have left?

    A. $75

    B. $90

    C. $105

    D. $120

16) A woman owns a dog walking business. If 3 workers can walk 9 dogs, how many dogs can 5 workers walk?

    A. 13

    B. 14

    C. 15

    D. 19

17) Will has been working on a report for 3 hours each day, 7 days a week for 2 weeks. How many minutes has will worked on his report?

    A. 6,364 Minutes

    B. 4,444 Minutes

    C. 2,520 Minutes

    D. 1,560 Minutes

18) A writer finishes 180 pages of his manuscript in 20 hours. How many pages is his average per hour?

    A. 18

    B. 14

    C. 12

    D. 9

19) Camille uses a 30% off coupon when buying a sweater that costs $50. If she also pays 5% sales tax on the purchase, how much does she pay?

    A. $35

    B. $36.75

    C. $39.95

    D. $47.17

20) I've got 34 quarts of milk and my family drinks 2 gallons of milk per week. How many weeks will that last us?

   A. 2 Weeks

   B. 2.5 Weeks

   C. 3.25 Weeks

   D. 4.25 Weeks

21) A floppy disk shows 937,036 bytes free and 739,352 bytes used. If you delete a file of size 652,159 bytes and create a new file of size 599,986 bytes, how many free bytes will the floppy disk have?

   A. 687,179

   B. 791,525

   C. 884,867

   D. 989,209

22) If one acre of forest contains 153 pine trees, how many pine trees are contained in 32 acres?

   A. 4,896

   B. 4,308

   C. 4,014

   D. 4,602

23) Five out of 30 students had to go to summer school. What is the ratio of students who did not have to go to summer school expressed, in its lowest terms?

   A. $\frac{5}{6}$

   B. $\frac{7}{8}$

   C. $\frac{3}{4}$

   D. $\frac{6}{7}$

24) A writer finishes 240 pages of his manuscript in 60 hours. How many pages is his average per hour?

   A. 15

   B. 12

   C. 6

   D. 4

25) Ava needs 1/5 of an ounce of salt to make 1 cup of dip for fries. How many cups of dip will she be able to make if she has 50 ounces of salt?

A. 35

C. 75

B. 200

D. 250

Test 1

Mathematics Knowledge

- o **25 questions**

- o **Total time for this section:** 22 Minutes

- o **Calculators are not allowed at the test.**

1) If a = 3, what is the value of b in this equation?

$$b = \frac{a^2}{3} + 3$$

A. 10　　　　　　　　　　　　　　　C. 6

B. 8　　　　　　　　　　　　　　　D. 4

2) The eighth root of 256 is:

A. 6　　　　　　　　　　　　　　　C. 8

B. 4　　　　　　　　　　　　　　　D. 2

3) A circle has a radius of 5 inches. What is its approximate area? (π = 3.14)

A. 90.7 square inches　　　　　　　C. 31.4 square inches

B. 78.5 square inches　　　　　　　D. 25 square inches

4) If $-8a = 64$, then $a =$ ___

A. –8　　　　　　　　　　　　　　　C. 16

B. 8　　　　　　　　　　　　　　　D. 0

5) In the following diagram what is the value of x?

A. 60∘

B. 90∘

C. 45∘

D. 15∘

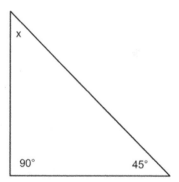

6) In the following right triangle, what is the value of x rounded to the nearest hundredth?

A. 23.24

B. 2.33

C. 10.29

D. 6.40

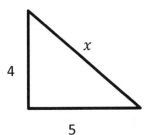

7) $(5x + 5)(2x + 6) = ?$

A. $5x + 6$

B. $10x^2 + 40x + 30$

C. $5x + 5x + 30$

D. $5x^2 + 5$

8) $5(a - 6) = 22$, what is the value of a?

A. 2.4

B. 10.4

C. 7

D. 11

9) If $3^{24} = 3^8 \times 3^x$, what is the value of x?

A. 2

B. 1.5

C. 3

D. 16

10) Which of the following is an obtuse angle?

A. 116°

B. 80°

C. 68°

D. 25°

11) Factor this expression: $x^2 + 5 - 6$

   A. $x^2(5 + 6)$ C. $(x + 6)(x - 1)$

   B. $x(x + 5 - 6)$ D. $(x + 6)(x - 6)$

12) Find the slope of the line running through the points (6, 7) and (5, 3).

   A. $\frac{1}{4}$ C. $- 4$

   B. 4   D. $- \frac{1}{4}$

13) What is the value of $\sqrt{100} \times \sqrt{36}$?

   A. 120   C. 60

   B. $\sqrt{136}$ D. $\sqrt{16}$

14) Which of the following is not equal to 5^2?

   A. the square of 5   C. 5 cubed

   B. 5 squared   D. 5 to the second power

15) The cube root of 2,197 is?

   A. 133   C. 6.5

   B. 13   D. 169

16) What is 952,710 in scientific notation?

   A. 95.271   C. 0.095271×10^6

   B. 9.5271×10^5 D. 0.95271

17) What's the area of the non-shaded part in the following figure?

A. 192

B. 152

C. 48

D. 40

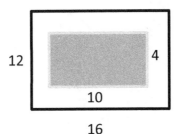

18) What's the square root of $16x^2$?

A. $4\sqrt{x}$

C. $4x$

B. $16x$

D. $\sqrt{4x}$

19) A square has one side with length 4.5 feet. The area of the square is:

A. 16 square feet

C. 20.25 square feet

B. 18.50 square feet

D. 25 square feet

20) A medium pizza has a diameter of 9 inches. What is its circumference?

A. 9π

C. 4.5π

B. $18\,\pi$

D. 3π

21) If a circle has a diameter of 2.5 feet, what is its circumference?

A. $2.5\,\pi$

C. $1.25\,\pi$

B. $5\,\pi$

D. $3.14\,\pi$

22) Which of the following sets of factors do both 76 and 20 have in common?

A. {1, 2, 4}

C. {0, 19, 38, 76}

B. {19, 38, 76}

D. {1, 2, 4, 96}

23) What is the circumference of a circle with center at point A if the distance from point X to Y is 22 feet? (π = 3.14)

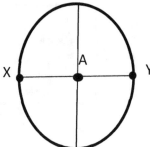

A. 74.13 Feet

B. 69.08 Feet

C. 36.26 Feet

D. 34.83 Feet

24) In the following diagram, the straight line is divided by one angled line at 115∘. What is the value of a.

A. 70∘

B. 90∘

C. 65∘

D. 180∘

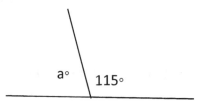

25) The volume of this box is:

A. 486 cm³

B. 586 cm³

C. 386 cm³

D. 243 cm³

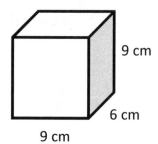

Test 2

Arithmetic Reasoning

- ○ **25 questions**

- ○ **Total time for this section:** 29 Minutes

- ○ **Calculators are not allowed at the test.**

1) If a box contains red and blue balls in ratio of 2 : 3, how many red balls are there if 90 blue balls are in the box?

A. 40   C. 80

B. 60   D. 30

2) A trash container, when empty, weighs 35 pounds. If this container is filled with a load of trash that weighs 240 pounds, what is the total weight of the container and its contents?

A. 224 Pounds   C. 285 Pounds

B. 275 Pounds   D. 325 Pounds

3) Which of the following is nine million, thirty–two thousand, six hundred two in standard form?

A. 932,620   C. 9,032,602

B. 932,602   D. 9,320,602

4) In a classroom of 44 students, 18 are male. About what percentage of the class is female?

A. 63%   C. 59%

B. 51%   D. 53%

5) The sum of 5 numbers is greater than 120 and less than 180. Which of the following could be the average (arithmetic mean) of the numbers?

A. 23   C. 37

B. 30   D. 42

6) How many ¼ pound paperback books together weigh 30 pounds?

A. 80   C. 105

B. 95   D. 120

7) A barista averages making 12 cups of coffee per hour. At this rate, how many hours will it take until she's made 960 cups of coffee?

A. 75   C. 85

B. 80   D. 90

Use the following table to answer question below.

DANIEL'S BIRD–WATCHING PROJECT	
DAY	NUMBER OF RAPTORS SEEN
Monday	?
Tuesday	9
Wednesday	14
Thursday	12
Friday	5
MEAN	10

8) The above table shows the data Daniel collects while watching birds for one week. How many raptors did Daniel see on Monday?

A. 10   C. 12

B. 11   D. 13

9) You just went on a blind date and wanted to impress the girl by adding an 20% tip to the total cost of the meal. Two entrees came to a total of $23.40. What is the 20% tip going to cost?

A. 1.91   C. 4.68

B. 3.91   D. 5.91

10) The drivers at G & G trucking must report the mileage on their trucks each week. The mileage reading of Ed's vehicle was 40,907 at the beginning of one week, and 41,053 at the end of the same week. What was the total number of miles driven by Ed that week?

A. 46 Miles

C. 146 Miles

B. 145 Miles

D. 1,046 Miles

11) Emily and Daniel have taken the same number of photos on their school trip. Emily has taken 5 times as many as photos as Claire and Daniel has taken 16 more photos than Claire. How many photos has Claire taken?

A. 4

C. 8

B. 6

D. 10

12) A baker uses 4 eggs to bake a cake. How many cakes will he be able to bake with 188 eggs?

A. 46

C. 48

B. 47

D. 49

13) Lily and Ella are in a pancake–eating contest. Lily can eat two pancakes per minute, while Ella can eat 2 ½ pancakes per minute. How many total pancakes can they eat in 5 minutes?

A. 9.5 Pancakes

C. 22.5 Pancakes

B. 29.5 Pancakes

D. 11.5 Pancakes

14) The distance between cities A and B is approximately 2,600 miles. If you drive an average of 68 miles per hour, how many hours will it take you to drive from city A to city B?

   A. approximately 41 hours

   B. approximately 38 hours

   C. approximately 29 hours

   D. approximately 27 hours

15) Henry is purchasing gifts for his family. So far he has purchased the following:
   - Four sweaters, each valued at $55
   - One computer game valued at $64
   - Two bracelets, each valued at $ 35

   Later, he returned one of the bracelets for a full refund and received a $4 rebate on the computer game. What is the total cost of the gifts after the refund and rebate?

   A. $247   C. $358

   B. $315   D. $368

16) The first four terms in a sequence are shown below. What is the fifth term in the sequence? {2, 4, 8, 14, ..}

   A. 14   C. 22

   B. 16   D. 34

17) How many square feet of tile is needed for a 18 foot x 18 foot room?

   A. 72 Square Feet   C. 324 Square Feet

   B. 108 Square Feet   D. 216 Square Feet

18) Convert 0.025 to a percent.

A. 0.03%

C. 2.50%

B. 0.25%

D. 25%

19) Find the average of the following numbers: 17, 13, 7, 21, 22

A. 16.5

C. 16

B. 17

D. 11

20) Liam is 64 years old, twice as old as Mason. How old is Mason?

A. 29 Years Old

C. 32 Years Old

B. 37 Years Old

D. 27 Years Old

21) If it takes three workers, working separately but at the same speed, 3 hours 25 minutes to complete a particular task, about how long will it take one worker, working at the same speed, to complete the same task alone?

A. 5 Hours 25 Minutes

C. 10 Hours

B. 9 Hours 45 Minutes

D. 10 Hours 15 Minutes

22) Daniel has been saving for a new car and finally has the money to make the deal. He put down $2,500 and will be paying $244 a month for 3.5 years. What is the total cost of his purchase?

A. $14,546

C. $6,998

B. $12,748

D. 8,948

23) A pizzeria caters parties and just finished baking 80 pizzas. The church youth group just picked up their order of 1/2 of the pizzas baked. A school graduation party is picking up 1/4 of the pizzas baked. A martial arts club stopped in and bought 1/10 of the pizzas baked. How many pizzas are left of the 80 that were baked?

    A. 5   C. 8

    B. 12   D. 10

24) A train must travel to a certain town in five days. The town is 4,500 miles away. How many miles must the train average each day to reach its destination?

    A. 900   C. 250

    B. 850   D. 550

25) Of the 2,400 videos available for rent at a certain video store, 600 are comedies. What percent of the videos are comedies?

    A. 18 ½ %

    B. 20%

    C. 22%

    D. 25%

Test 2

Mathematics Knowledge

- ○ **25 questions**

- ○ **Total time for this section:** 22 Minutes

- ○ **Calculators are not allowed at the test.**

1) If $9 + x \leq 18$, then $x \leq$?

    A. -9 C. 9

    B. 27   D. $27x$

2) What's the greatest common factor of the 18, 20 and 38?

    A. 5   C. 3

    B. 4   D. 2

3) Evaluate $5a + 25$, when $a = -5$

    A. -150 C. 50

    B. 30   D. 0

4) Which of the following is an acute angle?

    A. 218°   C. 118°

    B. 312°   D. 37°

5) What is 152.6588 rounded to the nearest hundredth?

    A. 152.65   C. 153

    B. 152.66   D. 152.659

6) What's the square root of $25x^2$?

    A. $5\sqrt{x}$ C. $5x$

    B. 25X   D. $\sqrt{5x}$

7) 12a + 20 = 140, a = ?

   A. 12   C. 14

   B. 10   D. 18

8) What is 5293.358821 rounded to the nearest 4 decimal places?

   A. 5294   C. 5293.3588

   B. 5293.359   D. 5293.36

9) $5 - 8 \div (4^2 \div 4) =$

   A. –6   C. 3

   B. – ½   D. 4

10) Simplify $\sqrt{32}$

   A. $2\sqrt{16}$ C. $4\sqrt{16}$

   B. $4\sqrt{2}$ D. $4\sqrt{3}$

11) The cube root of 729 is?

   A. $8\sqrt{2}$ C. 9

   B. 8   D. 888

12) A rectangle is cut in half to create two squares that each have an area of 36. What is the perimeter of the original rectangle?

   A. 18   C. 30

   B. 24   D. 36

13) What's the circumference of a circle that has a diameter of 18 m?

    A. 113.1 m

    B. 28.28 m

    C. 56.52 m

    D. 36 m

14) There are two pizza ovens in a restaurant. Oven 1 burns 4 times as many pizzas as oven 2. If the restaurant had a total of 25 burnt pizzas on Saturday, how many pizzas did oven 2 burn?

    A. 5

    B. 10

    C. 15

    D. 20

15) If x = 6, then $\frac{6^5}{x} =$

    A. 30

    B. 7,776

    C. 1,296

    D. 96

16) What is the value of 6! ?

    A. 720

    B. 18

    C. 14

    D. 12

17) What is one possible value of x in this equation? $x^2 - 2x - 15 = 0$

    A. 3

    B. 5

    C. –5

    D. 10

18) If a inches of rain falls in one minute, how many inches will fall in b hours?

    A. $60\frac{a}{b}$

    B. 60a

    C. 60b

    D. 60ab

19) Given the diagram of parallel lines, what is the value of a?

A. 248°

B. 68°

C. 58°

D. 112°

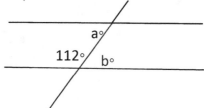

20) Given the diagram of parallelogram ABCD and the measure of A, what is the measurement of angle C?

A. 55°

B. 305°

C. 125°

D. 90°

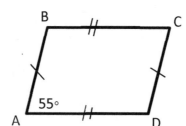

21) Solve for a: 8a – 5 = 11

A. 64

B. 0.45

C. 31.42

D. 2

22) What's the area of this trapezoid if the top width is 12, the bottom width is 24, and the height is 10?

A. 180

B. 72

C. 360

D. 140

23) What is the area of this trapezoid?

A. 52 cm^2

B. 140 cm^2

C. 191.2 cm^2

D. 390 cm^2

18 cm

12 cm

8 cm

14 cm

24) Which of the following pairs are complementary angles?

A. 150, 30

B. 120, 60

C. 70, −70

D. 70, 20

25) What is 56,800 in scientific notation?

A. 56.8 × 10^3

B. 5.68 × 10^4

C. 0.0568 × 10^6

D. 5.68

Test 3

Arithmetic Reasoning

- ○ **25 questions**

- ○ **Total time for this section:** 29 Minutes

- ○ **Calculators are not allowed at the test.**

1) Your local library branch has 30 employees and currently provides approximately 35 square feet per employee for a work station. The library just hired 2 new employees. To make room to add these 2 new people, each employee's work station will be reduced to about:

    A. 18.8 Square Feet   C. 32.8 Square Feet

    B. 20.8 Square Feet   D. 30.8 Square Feet

2) Which answer is equivalent to six to the fifth power?

    A. 0.0006   C. 1

    B. 60,000   D. 7,776

3) A woman weighs 135 pounds. She gains 15 pounds one month and 8 pounds the next month. What is her new weight?

    A. 152 Pounds   C. 158 Pounds

    B. 146 Pounds   D. 138 Pounds

4) Cory's average score after 6 tests is 90%. If he gets a 82% on his 7th test, what will his new average be?

    A. 88.86%   C. 87.8%

    B. 85.86%   D. 87.86%

5) Daniel bought a total of 8 packages of pens, and each package contained either 2 or 6 pens. If exactly 6 of the packages Daniel bought contained 6 pens, how many pens did Daniel buy?

    A. 16   C. 40

    B. 12   D. 54

6) Which of the following expressions is equal to 5^7 when $a = 5^4$?

    A. 10a   C. 14a

    B. 25a   D. 125a

7) Linda had $35.77 in her pocket. If shoes cost $ 8.25 a pair plus 10% tax. How many pairs can she buy?

    A. 3   C. 2

    B. 4   D. 5

8) An employee's rating on performance appraisals for the last three quarters were 92, 88 and 86. If the required yearly average to qualify for the promotion is 90, how much should the fourth quarter rating be?

    A. 90   C. 92

    B. 93   D. 94

9) A steak dinner at a restaurant costs $8.25. If a man buys a steak dinner for himself and 3 friends, what will the total cost be?

    A. $27   C. $33

    B. $17.01   D. $21.5

10) David's motorcycle stalled at the beach and he called the towing company. They charged him $3.45 per mile for the first 22 miles and then $4.25 per mile for each mile over 22. David was 26 miles from the motorcycle repair shop. How much was David's towing bill?

    A. $84.4   C. $90.9

    B. $71.9   D. $92.9

11) If Ella needed to buy 6 bottles of soda for a party in which 10 people attended, how many bottles of soda will she need to buy for a party in which 5 people are attending?

A. 3   C. 9

B. 6   D. 12

12) How much greater is the value of 5x + 8 than the value of $5x - 3$?

A. 7   C. 11

B. 9   D. 13

13) If a tutor earns $12 per hour, how much will she earn in a week if she reports on Monday and Thursdays, 9 am TO 12 noon and 1 PM to 6 PM, and Saturdays, 8 AM to 12 noon?

A. $210   C. $190

B. $220   D. $240

14) Amelia bought a pound of vegetables and used 3/5 of it to make a salad. How many ounces of vegetables are left after she makes the salad?

A. 1.2 Ounces   C. 4.4 Ounces

B. 2.2 Ounces   D. 6.4 Ounces

15) A florist has 516 flowers. How many full bouquets of 12 flowers can he make?

A. 42   C. 43

B. 44   D. 45

16) With what number must 5.674321 be multiplied in order to obtain the number 56743.21?

A. 100   C. 10,000

B. 1,000   D. 100,000

17) In a classroom of 50 students, 18 are male. What percentage of the class is female?

A. 59%   C. 63%

B. 51%   D. 64%

18) The bride and groom invited 220 guests for their wedding. 190 guests arrived. What percent of the guest list was not present?

A. 90%   C. 23.32%

B. 20%   D. 13.64%

19) Sofia takes $65 with her on a shopping trip to the mall. She spends $21 on new shoes and another $8 on lunch. How much money does she have left after these purchases?

A. $36   C. $50

B. $42   D. $26

20) If x is 25% percent of 250, what is x?

A. 35   C. 62.5

B. 95.5   D. 150

21) If 4 garbage trucks can collect the trash of 28 homes in a day. How many trucks are needed to collect in 70 houses?

A. 6   C. 10

B. 8   D. 20

22) ABC Corporation earned only $200,000 during the previous year, only two–third of the management's predicted income. How much earning did the management predict?

A. 20   C. 30,000

B. $300,000   D. 340,000

23) A car uses 15 gallons of gas to travel 450 miles. How many miles per gallon does the car get?

A. 28 miles per gallon

C. 30 miles per gallon

B. 32 miles per gallon

D. 34 miles per gallon

24) William keeps track of the length of each fish that he catches. Following are the lengths in inches of the fish that he caught one day: 13, 14, 9, 11, 9, 10, 18

What is the median fish length that William caught that day?

A. 11 Inches

C. 12 Inches

B. 9 Inches

D. 13 Inches

25) You're applying for a job that has a skills test requirement. In order to pass you need to get 35 of the 65 questions correct. About what percent of the questions do you need to answer correctly?

A. 42%

C. 22%

B. 54%

D. 34%

	Test 3
	Mathematics Knowledge

o **25 questions**

o **Total time for this section:** 22 Minutes

o **Calculators are not allowed at the test.**

1) Which of the following is not synonym for 8^2?

    A. the square of 8   C. 8 cubed

    B. 8 SQUARED   D. 8 to the second power

2) What's the greatest common factor of the 18 and 32?

    A. 11   C. 2

    B. 12   D. 4

3) What is 8923.2769 rounded to the nearest tenth?

    A. 8923.3   C. 8923

    B. 8923.277   D. 8923.27

4) What is the value of $\sqrt[3]{216}$?

    A. 3   C. 36

    B. 72   D. 6

5) If $x = 7$ what's the value of $6x^2 + 5x - 13$?

    A. 64   C. 416

    B. 316   D. 293

6) Which of the following is not a multiple of 5?

    A. 12   C. 15

    B. 30   D. 20

7) $25a - 15 = 220$, a = ?

   A. 8.8

   B. 8.2

   C. 9.4

   D. 16.6

8) What's the least common multiple (LCM) of 8 and 14?

   A. 8 and 14 have no common multiples

   B. 96

   C. 112

   D. 56

9) Line l passes through the point (−1, 2). Which of the following CANNOT be the equation of line l?

   A. $y = 1 - x$

   B. $y = x + 1$

   C. $x = -1$

   D. $y = x + 3$

10) The circumference of a circle is 30 cm. what is the approximate radius of the circle?

   A. 2.4 cm

   B. 4.8 cm

   C. 8.0 cm

   D. 9.5 cm

11) The cube of 8 is ___ .

   A. 512

   B. 64

   C. 80

   D. 24

12) $\dfrac{(15\ feet\ +7\ yards)}{4}$ = ___

    A. 9 Feet   C. 28 Feet

    B. 7 Feet   D. 4 Feet

13) $(p^4) \cdot (p^5)$ = ___

    A. P^{20} C. P^9

    B. $2P^9$ D. $2P^{20}$

14) In the diagram provided, what is the value of a?

    A. 225°

    B. 125°

    C. 55°

    D. 355°

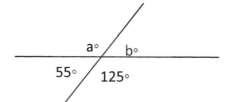

15) Use the diagram provided as a reference. If the length between point A and C is 95, and the length between point A and B is 36, what is the length between point B and C?

    A. 59

    B. 89

    C. 20

    D. 36

16) A rectangle is cut in half to create two squares that each have area of 9. What is the perimeter of the original rectangle?

    A. 12   C. 18

    B. 9   D. 21

17) If $(4.2 + 4.3 + 4.5)x = x$, then what is the value of x?

A. 0

C. 1

B. 1/10

D. 10

18) If $x + y = 12$, what is the value of $8x + 8y$?

A. 192

C. 104

B. 48

D. 96

19) Which of the following is an obtuse angle?

A. 60°

C. 135°

B. 85°

D. 28°

20) $(9^9)^7 = ?$

A. -9^3

C. 9^{63}

B. 9^2

D. 729^2

21) If $x = \frac{7}{9}$ then $\frac{1}{x} = ?$

A. $\frac{9}{7}$

C. 7

D. 9

B. $\frac{7}{9}$

22) In the circle seen below, if the two chords are perpendicular, what is the value of x?

A. 12.5

B. 25

C. 27

D. 30

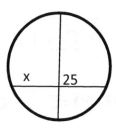

23) A given rectangle has an area of 24 square feet. Which of the following cannot be one of its side lengths?

A. 4

B. 8

C. 10

D. all of the above can be a side length.

24) The circumference of a circle is 40 cm. what is the approximate radius of the circle?

A. 2.4 cm

B. 4.8 cm

C. 6.37 cm

D. 12.76 cm

25) If n is a positive integer divisible by 8, and if n < 55, what is the greatest possible value of n?

A. 33

B. 40

C. 48

D. 53

Test 4

Arithmetic Reasoning

- ○ **25 questions**
- ○ **Total time for this section:** 29 Minutes
- ○ **Calculators are not allowed at the test.**

1) A diver can hold his breath for 1.5 minutes under water. After practicing for a week, he can hold his breath for 8% longer. How long will he be able to hold his breath after the first week of practice?

A. 2 Minutes

B. 0.62 Minutes

C. 1.62 Minutes

D. 1.062 Minutes

2) During a fund-raiser, each of the 45 members of a group sold candy bars. If each member sold an average of five candy bars, how many total bars did the group sell?

A. 21

B. 56

C. 195

D. 225

3) You work in a hospital and make $17.20 per hour for a 40 hour a week. If you work the night shift you get a slight differential added at $3.25 per hour worked during this shift. You just completed your first week and worked 25 hours on a regular shift and 5 hours on the night shift. What is the total amount you earned this week?

A. $527.9

B. $519.58

C. $532.5

D. $523.58

4) The supervisor of a small company is designing a new open office layout and thinks that employees will need 6 chairs for every 3 desks. How many office chairs will be needed for an office building with 60 desks?

A. 66 CHAIRS

B. 60 CHAIRS

C. 30 CHAIRS

D. 120 CHAIRS

5) On a map, the length of the road from town A to town B is measured to be 10 inches. On this map, ¼ inch represents an actual distance of 12 miles. What is the actual distance, in miles, from town A to town B along this road?

A. 340 Miles   C. 720 Miles

B. 480 Miles   D. 900 Miles

6) There are 55 rooms that need to be painted and only 8 painters available. If there are still 15 rooms unpainted by the end of the day, what is the average number of rooms that each painter has painted?

A. 6   C. 7

B. 5   D. 3

7) While preparing a dessert, Ella started by using 14 ounces of chocolate in her recipe. Later, she added 12 more ounces for flavor. What was the total amount of chocolate that Ella ended up using?

A. 1 Pound 4 Ounces   C. 1 Pound 8 Ounces

B. 1 Pound 6 Ounces   D. 1 Pound 10 Ounces

8) An auditorium that holds 550 people currently has 250 seated in it. What part of the auditorium is full?

A. 1/6   C. 5/11

B. 1/5   D. 5/9

9) Logan is scheduled to detail 33 cars this week. If he details 3 cars on Monday, how many must he detail each remaining day on average to finish by the end of the day on Saturday?

A. 6   C. 7

B. 5   D. 4

10) A man invested $220 in the stock market. During the first week, he lost $50. During the second week, he tripled his money. How much does he have at the end of the second week?

   A. $105

   B. $405

   C. $615

   D. $510

11) A piece of gauze 5 feet 5 inches long was divided in five equal parts. How long was each part?

   A. 1 Foot 7 Inches

   B. 15 Inches

   C. 11 Inches

   D. 13 Inches

12) With what number must 2.557159 be multiplied in order to obtain the number 25571.59?

   A. 100

   B. 10,000

   C. 1,000

   D. 100,000

13) The hour hand of a watch rotates 30 degrees every hour. How many complete rotations does the hour hand make in 4 days?

   A. 4

   B. 6

   C. 8

   D. 10

14) Sylvia is 7 years older than her sister Danna, and Danna is 5 years younger than their brother Jerry. If the sum of their ages is 72, how old is Danna?

   A. 18

   B. 22

   C. 26

   D. 20

15) Your uncle passed away and left you and your siblings his private property. He owned a 240 acre ranch and there are 4 of you. Each acre is valued at approximately $4,000. This means the monetary value for each sibling will be:

A. $240,000

C. 24,000

B. 80,000

D. 68,000

16) A woman owns a dog walking business. If 5 workers can walk 20 dogs, how many dogs can 8 workers walk?

A. 30

C. 32

B. 34

D. 36

17) What is five million, forty–five thousand, eight hundred three in numerals?

A. 545,803

C. 554,803

B. 5,045,803

D. 5,450,803

18) If you were making $6.00 per hour and got a raise to $6.75 per hour, what percentage increase was the raise?

A. 8%

C. 8%

B. 7%

D. 12.5%

19) You just drove 340 miles and it took you approximately 8 hours. How many miles per hour was your average speed?

A. About 44.5 Miles Per Hour

B. About 42.5 Miles Per Hour

C. About 46.5 Miles Per Hour

D. About 41.5 Miles Per Hour

20) Three people go to a restaurant. Their bill comes to $56.00. They decided to split the cost. One person pays $8.5, the next person pays 2 times that amount. How much will the third person have to pay?

   A. $36.50   C. $41.00

   B. $30.50   D. $44.00

21) Susan went to the Spa and got a manicure for $18.50, a pedicure for $32.25 and a massage for $40.50. She tipped a total of $5.20. What is the total Susan spent at the Spa?

   A. $96.45   C. $87.75

   B. $90.45   D. $71.25

22) If a store adds 30 chairs to its current inventory, the total number of chairs will be the same as three times the current inventory of chairs. If the manager wants to increase the current inventory by 60%, what will the new inventory of chairs be?

   A. 24   C. 60

   B. 30   D. 90

23) Which of the following is Not a factor of 60?

   A. 3   C. 9

   B. 6   D. 12

24) If x is 45% percent of 820, what is x?

   A. 185   C. 402

   B. 369   D. 720

25) If a rectangle measures 45 feet by 65 feet, what is its area?

    A. 2000   C. 2,925

    B. 220   D. 2,825

Test 4

Mathematics Knowledge

○ **25 questions**

○ **Total time for this section:** 22 Minutes

○ **Calculators are not allowed at the test.**

1) What is the value of 7!?

   A. 2,520

   B. 5,040

   C. 48

   D. 35

2) If $x + y = 6$, what is the value of 6x + 6y ?

   A. 36

   B. 60

   C. 12

   D. 26

3) Factor $x^2 - 25$

   A. $(x - 5)^2$

   B. $(x + 5)^2$

   C. $(x + 5)(x - 5)$

   D. $(x - 5)(x - 5)$

4) If $8x + 8y = 8$ find y when $x = 7$

   A. −2

   B. −6

   C. 7

   D. 6

5) What's the greatest common factor of 7 and 28?

   A. 1

   B. 2

   C. 14

   D. 7

6) What is the expanded equation from (2x + 6)(x − 4) = 0?

   A. $2x^2 + 2x - 24 = 0$

   B. $2x^2 - 2x - 24 = 0$

   C. $2x - 24x + 6 = 0$

   D. $8x^2 - 2x + 6 = 0$

7) Given that x = 0.6 and y = 6, what is the value of $2x^2(y + 4)$?

A. 7.2

C. 11.2

B. 8.2

D. 13.2

8) The base of a right triangle is 2 foot, and the interior angles are 45–45–90. What is its area?

A. 2 ft^2

C. 3.5 ft^2

B. 4 ft^2

D. 5.5 ft^2

9) Simplify the expression: $5x + 8.5x^2 = ?$

A. $11.5x^2$

B. $18x$

C. $64.5x$

D. $8.5x^2 + 5x$

10) The diagram shows parallelogram ABCD. What is the measure of angle A if angle D measures 123∘?

A. 123∘

B. 57∘

C. 180∘

D. 303∘

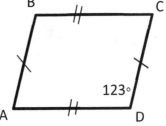

11) The average of 50, 35, 22 and 41 is ___ .

A. 37

C. 40

B. 34

D. 28

12) A cube has a volume of 729 cubic inches. What is the length one side of the cube?

A. 18 Inches   C. 9 Inches

B. 81 Inches   D. 162 Inches

13) Convert 319,000 to scientific notation.

A. 3.19×1000 C. 3.19×100

B. 3.19×10^{-5} D. 3.19×10^{5}

14) $\dfrac{(564 \cdot 2)}{6} = $ _____

A. 50   C. 10

B. 100   D. 188

15) Which of the following is a prime number: 11, 14, 6, or 18?

A. 11   C. 18

B. 14   D. 6

16) Which of these fractions is the largest?

$$\frac{5}{6}, \frac{1}{6}, \frac{2}{3}, \frac{1}{2}$$

A. $\dfrac{5}{6}$ C. $\dfrac{1}{6}$

B. $\dfrac{1}{2}$ D. $\dfrac{2}{3}$

17) A square box has 100-inch sides. What is its volume?

A. 10,000,000 Cubic Inches

B. 100,000 Cubic Inches

C. 1,000,000 Cubic Inches

D. 10,000

18) Solve for a: 8(a + 6) + 2 = 10

A. 36

B. 6

C. −5

D. −2

19) In the given diagram, the height is 9 cm. What is the area of the triangle?

A. 23 cm²

B. 46 cm²

C. 126 cm²

D. 252 cm²

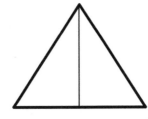

28 cm

20) If $9^{26} = 9^x \times 9^{12}$

A. 7

B. 9.5

C. 14

D. 24.5

21) Line l passes through the point (−2, 2). Which of the following cannot be the equation of line?

A. $y = -x$

B. $y = x + 1$

C. $x = -2$

D. $y = x + 4$

22) Which of the following is a composite number?

    A. 11   C. 3

    B. 13   D. 18

23) Which of the following is an obtuse angle?

    A. 89°   C. 143°

    B. 55°   D. 235°

24) A tube has a radius of 6 inches and a height of 10 inches. What it's approximate volume?

    A. 565.2 in^2 C. 1,130.4 in^2

    B. 188.4 in^2 D. 1,428.2 in^2

25) The cube root of 512 is?

    A. $4\sqrt{8}$

    B. 4

    C. 888

    D. 8

Test 5

Arithmetic Reasoning

- o **25 questions**

- o **Total time for this section:** 29 Minutes

- o **Calculators are not allowed at the test.**

1) Jean answered 24 out of the 30 questions correctly. What percentage of the total number of questions was she able to answer correctly?

   A. 7%

   B. 17%

   C. 70%

   D. 80%

2) Mario loaned Mia $1200 at a yearly interest rate of 5%. After one year what is the interest owned on this loan?

   A. $126

   B. $60

   C. $5

   D. $1260

3) What's the average of all the multiples of 5 from 1 to 40?

   A. 22.5

   B. 16.28

   C. 27.17

   D. 18

4) 57 passengers sit on my plane. How many trips will I have to make to transport 144 people 100 miles in 1 day?

   A. 4

   B. 5

   C. 3

   D. 2

5) Each sprinkler head on an athletic field sprays water at an average of 22 gallons per minute. If six sprinkler heads are following at the same time, how many gallons of water will be released in 10 minutes?

   A. 132 Gallons

   B. 1,320 Gallons

   C. 220 Gallons

   D. 2,690 Gallons

6) If $5y + 3y + 4y = -12$, then what is the value of y?

   A. −1

   B. 0

   C. 1

   D. 2

7) The total of the ages of Jeffrey, Edgar and Erik is 95 years. What was the total of their ages three years ago?

   A. 86 Years

   B. 87 Years

   C. 89 Years

   D. 92 Years

8) Matias is having a birthday party for his son and is serving orange juice to the 12 children in attendance. If Matias has 2 liter of orange juice and wants to divide it equally among the children, how many liters does each child get?

   A. $\frac{1}{3}$ of litter

   B. $\frac{1}{5}$ of litter

   C. $\frac{1}{7}$ of litter

   D. $\frac{1}{6}$ of litter

9) What's the next number in the series {20, 17, 14, 11, ?}

   A. 8

   B. 12

   C. 6

   D. 15

10) A man goes to a grocery store. He buys a carton of milk for $5.20, a dozen eggs for $1.80, and a pound of ground beef for $3. If the tax is 8%, what will his total bill be?

   A. $8.99

   B. $8.014

   C. $10.80

   D. $10.08

11) Find the average of the following numbers: 12, 28, 18, 24, 16, 14

   A. 27.37

   B. 19.37

   C. 18.67

   D. 10.67

12) The sum of 8 numbers is greater than 240 and less than 320. Which of the following could be the average (arithmetic mean) of the numbers?

   A. 30

   B. 35

   C. 40

   D. 45

13) A barista averages making 15 coffee per hour. At this rate, how many hours will it take until she's made 855 coffee?

   A. 52

   B. 57

   C. 63

   D. 68

14) A steak dinner at a restaurant costs $4.75. If a man buys a steak dinner for himself and 5 friends, what will the total cost be?

   A. $23.75

   B. $12.51

   C. $28.5

   D. $17

15) Convert 0.087 to a percent.

   A. 0.09%

   B. 0.87%

   C. 8.70%

   D. 87%

16) If 6 garbage trucks can collect the trash of 36 homes in a day. How many trucks are needed to collect in 180 houses?

   A. 18

   B. 19

   C. 15

   D. 30

17) A philanthropist donates 2/5 of his monthly income for the construction of the nearby church. How much percentage of his monthly income is he donating?

   A. 40%

   B. 4%

   C. 14%

   D. 400%

18) An item in the store originally priced at $300 was marked down 20%. What is the final sale price of the item?

    A. $240

    B. $204

    C. $195

    D. $200

19) Ellis just got hired for on–the–road sales and will travel about 2,500 miles a week during a 80 hour work week. If the time spent raveling is $\frac{3}{5}$ of his week, how many hours a week will he be on the road?

    A. Ellis spends about 34 hours of his 80 hour work week on the road.

    B. Ellis spends about 48 hours of his 80 hour work week on the road.

    C. Ellis spends about 43 hours of his 80 hour work week on the road.

    D. Ellis spends about 40 hours of his 80 hour work week on the road.

20) A triathlon course includes a 800m swim, a 55.7km bike ride, and a 3.65km run. What is the total length of the race course?

    A. 922 km

    B. 67.35 km

    C. 922 m

    D. 60.15 km

21) How many 18–passenger vans will it take to drive all 48 players and coaches on the football team to the away game?

    A. 4

    B. 3

    C. 2

    D. 1

22) Which of the following is an integer?

    A. −7

    B. 2/5

    C. 0.8

    D. $\sqrt{37}$

23) Which of the following is equivalent to $7 \times 7 \times 7 \times 7$?

    A. $\sqrt{7} \times 7$

    B. 7^4

    C. $7 \div 7$

    D. 7,000

24) The absolute value of 8 plus a number equals 22. Which of the following options is the correct solution set for the above equation?

    A. {−14, 14}

    B. {−20, 20}

    C. {−14, 30}

    D. {−30, 14)

25) A 3 ton truck is taxed at a rate of $0.18 per pound. How much is owed in taxes?

    A. $0.18

    B. $2,160

    C. $1,080

    D. $180

Test 5

Mathematics Knowledge

- o **25 questions**

- o **Total time for this section:** 22 Minutes

- o **Calculators are not allowed at the test.**

1) A pizza maker has x pounds of flour to make pizzas. After he has used 55 pounds of flour, how much flour is left? The expression that correctly represents the quantity of flour left is:

A. $55 + x$

B. $\dfrac{x}{55}$

C. $55 - x$

D. $x - 55$

2) What is the value of $5!$?

A. 120

B. 15

C. 12

D. 10

3) If $a^5 + b^5 = a^5 + c^5$, then b = ___

A. c

B. A

C. $B^4 - A^4$

D. $a^4 - b^4$

4) $(x + 8)(3x + 6) =$ ____

A. $3x^2 + 30x + 14$

B. $3x^2 + 14x + 48$

C. $3x^2 + 30x + 30$

D. $3x^2 + 30x + 48$

5) In the diagram of parallel lines provided, what is the measure of angle a?

A. 60∘

B. 120∘

C. 70∘

D. 180∘

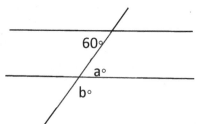

6) Given the diagram of parallelogram ABCD and the measure of A, what is the measurement of angle of C?

A. 90∘

B. 214∘

C. 107∘

D. 73∘

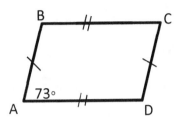

7) If $7^{14} = 7^5 \times 7^x$

    A. 3

    B. 5.5

    C. 9

    D. 12.5

8) What's the circumference of a circle that has a radius of 6 cm?

    A. 25.13 cm^2

    B. 37.7 cm^2

    C. 31.42 cm^2

    D. 44 cm^2

9) Find the diagonal of a square whose area is 25.

    A. 5

    B. $25\sqrt{2}$

    C. 2

    D. $5\sqrt{2}$

10) If a square box has a volume of 125 in^3 what's the perimeter of each face of the box?

    A. 20 INCHES

    B. 10 INCHES

    C. 24 INCHES

    D. 5 INCHES

11) Which of the following is the symbol for large than?

    A. ≥   C. <

    B. >   D. ≤

12) What is 7502.467355 rounded to the nearest 1 decimal places?

    A. 7502.467   C. 7502.5

    B. 7502.4673   D. 7502.47

13) What's the reciprocal of $\frac{x^3}{16}$?

    A. $\frac{16}{x^3} - 1$ C. $\frac{16}{x^3} + 1$

    B. $\frac{48}{x^3}$ D. $\frac{16}{x^3}$

14) Which of the following is a composite number?

    A. 24   C. 13

    B. 11   D. 19

15) Find the diagonal of a square whose area is 49.

    A. 7   C. 3

    B. $7\sqrt{2}$ D. $49\sqrt{2}$

16) Solve for the value of y in the following system of equations:
$4x - 6y = -8$, $8x + 4y = 16$

    A. −0.2   C. 4

    B. 2   D. −2.75

17) You are told that one of the internal angles of a parallelogram is 68 degrees. This means at least one other angle must be equal to:

A. 86 Degrees

C. 12 Degrees

B. 123 Degrees

D. 68 Degrees

18) Calculate the area of a parallelogram with a base of 2 feet and height of 2.4 feet.

A. 2.8 Square Feet

C. 4.8 Square Feet

B. 4.2 Square Feet

D. 4.0 Square Feet

19) What is one possible value of x in this equation? $x^2 - 9x + 18 = 0$

A. 3

C. −5

B. 5

D. 10

20) Calculate the area of the circle shown in the figure.

A. 64 π square cm

B. 64 π square cm

C. 100.5 π square cm

D. 201.1 π square cm

21) What is the area of an isosceles right triangle that has one leg that measures 4 cm?

A. 8 cm²

C. 6 $\sqrt{2}$ cm²

B. 36 cm²

D. 72 cm²

22) If $\dfrac{0.09}{a} = 0.01$, then a = _____ .

A. 0.09

C. 90

B. 0.9

D. 9

23) What is 796,000 in scientific notation?

A. 7^{96}

C. 7.96

B. 7.96×10^5

D. 79.6

24) If a is a positive number and b is a negative number which of these statements must be true?

A. $-ab$ is a negative number

B. $ab > 0$

C. ab^2 is a positive number

D. ab is a positive number

25) What is the value of $\sqrt{49} \times \sqrt{25}$?

A. 44

C. 35

B. $\sqrt{24}$

D. $\sqrt{17.5}$

Test 6

Arithmetic Reasoning

- ○ **25 questions**

- ○ **Total time for this section:** 29 Minutes

- ○ **Calculators are not allowed at the test.**

1) many minutes has Will worked on his report?

   A. 42

   B. 84

   C. 2,520

   D. 5,040

2) James is driving to visit his mother, who lives 340 miles away. How long will the drive be, round–trip, if James drives at an average speed of 50 mph?

   A. 135 minutes

   B. 310 minutes

   C. 741 minutes

   D. 816 minutes

3) In a classroom of 60 students, 42 are female. What percentage of the class is male?

   A. 34%

   B. 22%

   C. 30%

   D. 26%

4) You are asked to chart the temperature during a 6-hour period to give the average. These are your results:

   7 am: 7 degrees

   8 am: 9 degrees

   9 am: 22 degrees

   10 am: 28 degrees

   11 am: 28 degrees

   12 pm: 30 degrees

   What is the average temperature?

   A. 32.67

   B. 24.67

   C. 20.67

   D. 18.27

5) During the last week of track training, Emma achieves the following times in seconds: 66, 57, 54, 64, 57, and 59. Her three best times this week (least times) are averaged for her final score on the course. What is her final score?

A. 56 seconds

C. 59 seconds

B. 57 seconds

D. 61 seconds

6) How many square feet of tile is needed for a 15 feet x 15 feet room?

A. 225 square feet

C. 112 square feet

B. 118.5 square feet

D. 60 square feet

7) With what number must 1.303572 be multiplied in order to obtain the number 1303.572?

A. 100

C. 10,000

B. 1,000

D. 100,000

8) Which of the following is NOT a factor of 50?

A. 5

C. 10

B. 2

D. 15

9) Emma is working in a hospital supply room and makes $25.00 an hour. The union negotiates a new contract giving each employee a 4% cost of living raise. What is Emma's new hourly rate?

A. $26 an hour

C. $30 an hour

B. $28 an hour

D. $31.50 an hour

10) Emily and Lucas have taken the same number of photos on their school trip. Emily has taken 4 times as many photos as Mia. Lucas has taken 21 more photos than Mia. How many photos has Mia taken?

A. 7

B. 9

C. 11

D. 13

11) Will has been working on a report for 5 hours each day, 6 days a week for 2 weeks. How many minutes has Will worked on his report?

A. 7,444 minutes

B. 5,524 minutes

C. 3,600 minutes

D. 2,640 minutes

12) Find the average of the following numbers: 22, 34, 16, 20

A. 23

B. 35

C. 30

D. 23.3

13) A mobile classroom is a rectangular block that is 90 feet by 30 feet in length and width respectively. If a student walks around the block once, how many yards does the student cover?

A. 2,700 yards

B. 240 yards

C. 120 yards

D. 60 yards

14) What is the distance in miles of a trip that takes 2.1 hours at an average speed of 16.2 miles per hour? (Round your answer to a whole number)

A. 44 miles

B. 34 miles

C. 30 miles

D. 18 miles

15) The sum of 6 numbers is greater than 120 and less than 180. Which of the following could be the average (arithmetic mean) of the numbers?

A. 20

C. 30

B. 26

D. 34

16) A barista averages making 15 coffees per hour. At this rate, how many hours will it take until she's made 1,500 coffees?

A. 95 hours

C. 100 hours

B. 90 hours

D. 105 hours

17) There are 120 rooms that need to be painted and only 12 painters available. If there are still 12 rooms unpainted by the end of the day, what is the average number of rooms that each painter has painted?

A. 9

C. 14

B. 12

D. 16

18) Nicole was making $7.50 per hour and got a raise to $7.75 per hour. What percentage increase was Nicole's raise?

A. 2%

C. 3.33%

B. 1.67%

D. 6.66%

19) An architect's floor plan uses ½ inch to represent one mile. What is the actual distance represented by 4 ½ inches?

A. 9 miles

C. 7 miles

B. 8 miles

D. 6 miles

20) A snack machine accepts only quarters. Candy bars cost 25¢, a package of peanuts costs 75¢, and a can of cola costs 50¢. How many quarters are needed to buy two Candy bars, one package of peanuts, and one can of cola?

A. 8 quarters

C. 6 quarters

B. 7 quarters

D. 5 quarters

21) The hour hand of a watch rotates 30 degrees every hour. How many complete rotations does the hour hand make in 8 days?

A. 12

C. 16

B. 14

D. 18

22) What is the product of the square root of 81 and the square root of 25?

A. 2,025

C. 25

B. 15

D. 45

23) If $2y + 4y + 2y = -24$, then what is the value of y?

A. −3

C. −1

B. −2

D. 0

24) A bread recipe calls for $2\frac{2}{3}$ cups of flour. If you only have $1\frac{5}{6}$ cups of flour, how much more flour is needed?

A. 1

C. 2

B. $\frac{1}{2}$

D. $\frac{5}{6}$

25) Convert 0.023 to a percent.

   A. 0.2%

   B. 0.23%

   C. 2.30%

   D. 23%

Test 6

Mathematics Knowledge

- ○ **25 questions**

- ○ **Total time for this section:** 22 Minutes

- ○ **Calculators are not allowed at the test.**

1) $(x + 7)(x + 5) = ?$

    A. $x^2 + 12x + 12$ C. $x^2 + 35x + 12$

    B. $2x + 12x + 12$ D. $x^2 + 12x + 35$

2) Convert 670,000 to scientific notation.

    A. 6.70×1000 C. 6.70×100

    B. 6.70×10^{-5} D. 6.7×10^5

3) What is the perimeter of the triangle in the provided diagram?

    A. 15,625

    B. 625

    C. 75

    D. 25

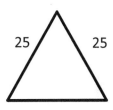

4) If x is a positive integer divisible by 6, and $x < 60$, what is the greatest possible value of x?

    A. 54   C. 36

    B. 48   D. 59

5) There are two pizza ovens in a restaurant. Oven 1 burns four times as many pizzas as oven 2. If the restaurant had a total of 15 burnt pizzas on Saturday, how many pizzas did oven 2 burn?

    A. 3   C. 9

    B. 6   D. 12

6) Which of the following is an obtuse angle?

    A. 56°

    B. 72°

    C. 123°

    D. 211°

7) $7^7 \times 7^8 = ?$

    A. 7^{56}

    B. $7^{0.89}$

    C. 7^{15}

    D. 1^7

8) What is 5231.48245 rounded to the nearest tenth?

    A. 5231.482

    B. 5231.5

    C. 5231

    D. 5231.48

9) The cube root of 512 is?

    A. 8

    B. 88

    C. 888

    D. 134,217,728

10) A circle has a diameter of 16 inches. What is its approximate area? ($\pi = 3.14$)

    A. 200.96

    B. 100.48

    C. 64.00

    D. 12.56

11) Which of the following is the correct calculation for 7!?

    A. $7 \times 6 \times 5 \times 4 \times 3 \times 2 \times 1$

    B. $1 \times 2 \times 3 \times 4 \times 5 \times 6$

    C. $0 \times 1 \times 2 \times 3 \times 4 \times 5 \times 6 \times 7$

    D. $1 \times 2 \times 3 \times 4 \times 5 \times 6 \times 7 \times 8$

12) The equation of a line is given as: $y = 5x - 3$. Which of the following points does not lie on the line?

    A. (1, 2)   C. (3, 18)

    B. (−2, −13)   D. (2, 7)

13) How long is the line segment shown on the number line below?

    A. −9   C. 8

    B. −8   D. 9

-10 -9 -8 -7 -6 -5 -4 -3 -2 -1 0 1 2 3 4 5 6 7 8 9 10

14) What is the distance between the points (1, 3) and (−2, 7)?

    A. 3   C. 5

    B. 4   D. 6

15) $x^2 - 81 = 0$, x could be:

    A. 6   C. 12

    B. 9   D. 15

16) A rectangular plot of land is measured to be 160 feet by 200 feet. Its total area is:

    A. 32,000 square feet   C. 3,200 square feet

    B. 4,404 square feet   D. 2,040 square feet

17) With what number must 2.103119 be multiplied in order to obtain the number 21,031.19?

    A. 100   C. 10,000

    B. 1,000   D. 100,000

18) Which of the following is NOT a factor of 50?

A. 5

B. 10

C. 2

D. 100

19) The sum of 4 numbers is greater than 320 and less than 360. Which of the following could be the average (arithmetic mean) of the numbers?

A. 80

B. 85

C. 90

D. 95

20) One fourth the cube of 4 is:

A. 25

B. 16

C. 32

D. 8

21) What is the sum of the prime numbers in the following list of numbers?

14, 12, 11, 16, 13, 20, 19, 36, 30

A. 26

B. 37

C. 43

D. 32

22) Convert 25% to a fraction.

A. 1/2

B. 2/3

C. 1/4

D. 3/4

23) The supplement angle of a 45° angle is:

A. 135°

B. 105°

C. 90°

D. 35°

24) 20% of 50 is:

    A. 30

    B. 25

    C. 20

    D. 10

25) Simplify: $5(2x^6)^3$.

    A. $10x^9$

    B. $10x^{18}$

    C. $40x^{18}$

    D. $40x^9$

Test 7

Arithmetic Reasoning

o **25 questions**

o **Total time for this section:** 29 Minutes

o **Calculators are not allowed at the test.**

1) A waiter earns $12.00 an hour when he serves in a shift. If he works a 40 hour shift in a week, his total earnings for the week would be ___ .

   A. $120.00

   B. $480.00

   C. $48.00

   D. $100.00

2) A company pays its writer $4 for every 400 words written. How much will a writer earn for an article with 960 words?

   A. $11

   B. $5.60

   C. $9.60

   D. $8.70

3) Seven girls in a club are comparing their weekly allowance. Ella and Avery each receives $12 a week. Ava and Emily each receives $9 a week. Sofia, Chloe, and Emma each receives $14 a week. What is the average amount of allowance given among these 7 girls?

   A. $11.71

   B. $12.88

   C. $12.00

   D. $10.20

4) Two monorails depart their respective stations at the same time. The monorail headed south is traveling 43 miles an hour, while the monorail headed north is traveling 64 miles per hour. Given this information, how many miles will both monorails travel in 6 hours?

   A. 732 Miles

   B. 670 Miles

   C. 627 Miles

   D. 642 Miles

5) Charlotte is 46 years old, twice as old as Avery. How old is Avery?

   A. 23 Years Old

   B. 28 Years Old

   C. 20 Years Old

   D. 18 Years Old

6) David is just beginning a computer consulting firm and has purchased the following equipment:

- Three telephone sets, each costing $125
- Two computers, each costing $1,300
- Two computer monitors, each costing $950
- One printer costing $600

David is reviewing his finances. What should he write as the total value of the equipment he has purchased so far?

A. $3,025   C. $5,525

B. $5,475   D. $6,525

7) Alice is working in a hospital supply room and make $58.00 an hour. The union negotiates a new contract giving each employee a 5% cost of living raise. What is Alice's new hourly rate?

A. $60.90 an hour   C. $62.76 an hour

B. $61.00 an hour   D. $64.00 an hour

8) What is the value of $\frac{4}{13} \times \frac{3}{13}$?

A. $\frac{7}{26}$ C. $1\frac{7}{26}$

B. $1\frac{12}{121}$ D. $\frac{12}{169}$

9) Elizabeth wants to give her granddaughter a present of $420 saved in a bank account. If Ellie puts $300 in the account which earns an 5% annual simple interest rate, how long must she wait before the account is worth $420?

A. 5 Years 6 Months   C. 6 Years 11 Months

B. 7 Years 5 Months   D. 8 Years

10) If x does not equal zero, and $xy = \frac{y}{5}$, what is the value of x?

    A. 1/9   C. 1/4

    B. 1/5   D. 1/3

11) Cory's average score after 5 tests is 82. If he gets a 67 on his 6th test, what will his new average be?

    A. 79.5   C. 77.4

    B. 80.5   D. 79.4

12) A cross country coach just announced that, by the end of the quarter, all students should run the 10 miles in less than 70 minutes. There are 15 students in the group.

    6 of them ran 10 miles in 1 hour + 5 minutes

    4 of them ran 10 miles in 1 hour + 3 minutes

    2 of them ran 10 miles in 1 hour + 9 minutes

    3 of them ran 10 miles in 1 hour and 10 minutes.

    How many students made the goal?

    A. 12   C. 10

    B. 4   D. 3

13) A pole is 8 feet long. Its shadow is 6 feet long. If you draw a line from the tip of the pole to the tip of the shadow, how long will the line be?

    A. 18 Feet   C. 20 Feet

    B. 12 Feet   D. 10 Feet

14) If the ratio of home fans to visiting fans in a crowd is 3:2 and all 25,000 seats in a stadium are filled, how many visiting fans are in attendance?

A. 100,000

C. 1,000

B. 100

D. 10,000

15) Chad's recipe for chocolate chip cookies makes 32 servings of 425 calories each. Chad decided to make 135% of the amount in the recipe rather than the usual 100%. Approximately how many calories are in Chad's batch of cookies?

A. 16,060

C. 14,560

B. 18,360

D. 11,260

16) On a map, the length of the road from Town F to Town G is measured to be 22 inches. On this map, 1/3 inch represents an actual distance of 15 miles. What is the actual distance, in miles, from Town F to Town G along this road?

A. 850 Miles

C. 14,560 Miles

B. 990 Miles

D. 11,260 Miles

17) If an object travels at 0.3 cm per second, how many meters does it travel in 4 hours?

A. 21.5

C. 2

B. 88.2

D. 43.2

18) How many 1/2 pound paperback books together weigh 30 pounds?

A. 80

C. 105

B. 60

D. 120

19) Five out of 25 students had to go to summer school. What is the ratio of students who did not have to go to summer school to total number of students, expressed in its lowest terms?

A. $\dfrac{4}{5}$ C. $\dfrac{7}{8}$

B. $\dfrac{2}{3}$ D. $\dfrac{5}{6}$

20) If a mayor is elected with 70% of the votes cast and 78% of the town's 22,000 voters cast a vote, how many votes did the mayor receive?

A. 12,012   C. 15,200

B. 8,975   D. 12,245

21) Carl works for a newspaper and wants to write 20 short articles this week. He writes 6 articles on Monday. If Carl works Monday through Friday, how many articles should he write each day for the rest of the week in order to reach his goal?

A. 4.5   C. 2.5

B. 3.5   D. 1.5

22) While at work, Emma checks her email once every 40 minutes. In 8−hour, how many times does she check her email?

A. 10 Times   C. 13 Times

B. 11 Times   D. 12 Times

23) A company pays its writer $2 for every 300 words written. How much will a writer earn for an article with 1200 words?

A. $10   C. $8

B. $4   D. $7

24) Levi keeps track of the length of each fish that he catches. Following are the lengths in inches of the fish that he caught one day: 22, 21, 19, 14, 17, 13, 14

What is the average fish length that Levi caught that day?

A. 15 Inches   C. 18 Inches

B. 17 Inches   D. 19 Inches

25) What's the next number in the series {29, 24, 19, 14, ?}

A. 14   C. 4

B. 9   D. 19

Test 7

Mathematics Knowledge

o **25 questions**

o **Total time for this section:** 22 Minutes

o **Calculators are not allowed at the test.**

1) If $a = 8$, what is the value of b in this equation?

$$b = \frac{a^2}{4} + 4$$

A. 24   C. 20

B. 22   D. 28

2) If a rectangular swimming pool has a perimeter of 112 feet and is 22 feet wide, what is its area?

A. 1,496   C. 2,464

B. 90   D. 748

3) $(x^8)^2$

A. $2x^8$ C. x^{16}

B. x^{28} D. x^{10}

4) In the diagram provided, what is the value of b?

A. 180°

B. 120°

C. 135°

D. 45°

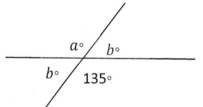

5) The volume of this box is:

A. 30 cm³

B. 42 cm³

C. 35 cm³

D. 210 cm³

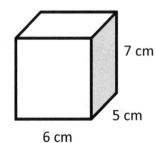

6) Which of the following is a prime number?

   A. 10

   B. 8

   C. 11

   D. 93

7) If $a = 8$ what's the value of $4a^2 + 3a + 10$?

   A. 166

   B. 216

   C. 290

   D. 276

8) Which of the following pairs are supplementary angles?

   A. 27°, −27°

   B. 49°, 61°

   C. 139°, 161°

   D. 68°, 112°

9) A circle has a diameter of 8 inches. What is its approximate circumference?

   A. 6.28 Inches

   B. 25.12 Inches

   C. 34.85 Inches

   D. 35.12 Inches

10) What is 781,200 in scientific notation?

   A. 78.12

   B. 7.812×10^5

   C. 0.07812×10^6

   D. 0.7812

11) Simplify $\sqrt{27}$

   A. $3\sqrt{27}$

   B. $2\sqrt{27}$

   C. $3\sqrt{3}$

   D. $2\sqrt{3}$

12) If $x = \frac{5}{7}$ then $\frac{1}{x} = $?

    A. $\frac{7}{5}$ C. 5

                                              D. 7

    B. $\frac{5}{7}$

13) If $6.5 < x \leq 9.0$, then x cannot be equal to:

    A. 6.5   C. 7.2

    B. 9   D. 7.5

14) Which of the following angles is obtuse?

    A. 24 Degrees   C. 190 Degrees

    B. 54 Degrees   D. 145 Degrees

15) Calculate the area of the trapezoid in the figure in square feet.

    A. 4.5 ft^2

    B. 6.5 ft^2

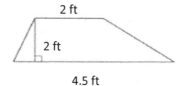

    C. 13 ft^2

    D. 26 ft^2

16) Which of the following is equivalent to x^{-6}?

    A. $-6x$ C. $-x^6$

    B. $\frac{1}{x^6}$ D. $\frac{x}{6}$

17) What percentage of 500 is, 350?

    A. 70%   C. 55%

    B. 35%   D. 80%

18) The product a number and its square is 343. What is the number?

    A. 171.5   C. 14

    B. 7   D. 85.75

19) Two ships leave from a port. Ship a sails west for 200 miles, and ship b sails north for 100 miles. How far apart are the ships after their trips?

    A. 150 Miles   C. 223.6 Miles

    B. 200 Miles   D. 300 Miles

20) What is the value of the expression $\frac{x-y}{z-x}$ when $x = 10$, $y = 5$ and $z = 15$?

    A. 1   C. 2

    B. −2   D. −1

21) What is the perimeter of the following regular polygon?

    A. 18 cm

    B. 22 cm   5 cm

    C. 30 cm

    D. 38 cm

22) The mean of, 3, 6, 2, 10, and x is 6. What is the value of x?

    A. 8   C. 10

    B. 9   D. 11

23) The measure of angle A is 50°. What is the measure of the complement of angle A?

    A. 40°   C. 310°

    B. 130°   D. 50°

24) $(8 - 2)! = ?$

    A. 6   C. 30

    B. 720   D. 120

25) What is the length b in the following right triangle?

    A. 5

    B. 6

    C. 7

    D. 8

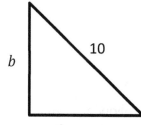

Test 8

Arithmetic Reasoning

- ○ **25 questions**

- ○ **Total time for this section:** 29 Minutes

- ○ **Calculators are not allowed at the test.**

1) Jason has been working on a report for 3 hours each day, 6 days a week for 4 weeks. How many minutes has Jason worked on his report?

   A. 4,320 Minutes

   B. 4,444 Minutes

   C. 2,520 Minutes

   D. 1,560 Minutes

2) Find the average of the following numbers: 32, 30, 25, 20, 14, 18

   A. 32.27

   B. 24.27

   C. 15.17

   D. 23.17

3) A classroom is a rectangular block that is 10 feet by 50 feet in length and width respectively. If a student walks around the block once, how many yards does the student cover?

   A. 40 Yards

   B. 35 Yards

   C. 45 Yards

   D. 30 Yards

4) What's the next number in the series {32, 29, 26, 23, ?}

   A. 15

   B. 20

   C. 17

   D. 13

5) What is the distance in miles of a trip that takes 4.2 hours at an average speed of 26.4 miles per hour?

   A. 55.44 Miles

   B. 112.0 Miles

   C. 110.88 Miles

   D. 109.80 Miles

6) How many 1/5 pound paperback books together weigh 20 pounds?

   A. 60

   B. 50

   C. 90

   D. 100

7) What is the product of the square root of 256 and the square root of 121?

A. 30,976

C. 196

B. 176

D. 88

8) The sum of 9 numbers is greater than 450 and less than 540. Which of the following could be the average (arithmetic mean) of the numbers?

A. 40

C. 50

B. 45

D. 55

9) A landscaping company charges 8 cents per square foot to apply fertilizer. How much would it cost for them to fertilize a 22 ft. x 32 ft. lawn?

A. $7.04

C. $56.32

B. $704

D. $5632

10) If Jill needed to buy 7 bottles of soda for a party in which 14 people attended, how many bottles of soda will she need to buy for a party in which 6 people are attending?

A. 3

C. 7

B. 5

D. 9

11) Convert 0.014 to a fraction.

A. $\dfrac{7}{100}$

C. $\dfrac{7}{500}$

B. $\dfrac{14}{100}$

D. $2\dfrac{3}{50}$

12) What is the result when 9 is added to the product of 4 and 6?

    A. 30   C. 19

    B. 20   D. 33

13) 26 members of a bridal party need transported to a wedding reception but there are only five 5–passenger taxis available to take them. How many will need to find other transportation?

    A. 0   C. 2

    B. 1   D. 3

14) If doughnuts are usually 75 cents each, but there is a sale on Friday advertising them as 1 dozen + ½ dozen for $10.80, what is the new cost for just one?

    A. 60 Cents   C. 65 Cents

    B. 50 Cents   D. 55 Cents

15) Which answer is equivalent to four to the fifth power?

    A. 0.00004   C. 1

    B. 400,000   D. 1,024

16) A tiger in a zoo has consumed 56 pounds of food in 7 days. If the tiger continues to eat at the same rate, in how many more days will its total food consumption be 77 pounds?

    A. 5   C. 6

    B. 3   D. 1

17) Solve the following equation: $|9 - (12 \div |\, 2 - 5\, |)|$

    A. 9   C. 5

    B. –6   D. –5

18) A menswear store is having a sale. Buy one dress shirt at full price or get two shirts for 15% off. If Sam buys two shirts, each with a regular price of $40.00, how much money will he save?

A. $6.00

C. $24.00

B. $12.00

D. $68.00

19) A writer finishes 640 pages of his manuscript in 80 hours. How many pages is his average?

A. 19

C. 10

B. 16

D. 8

20) 92 passengers fit on my plane. How many trips will I have to make to transport 460 people 100 miles in 1 day?

A. 4

C. 3

B. 5

D. 6

21) Which of the following is equal to the square root of 72?

A. $2\sqrt{6}$

C. $6\sqrt{2}$

B. $36\sqrt{2}$

D. $12\sqrt{6}$

22) You just went on a blind date and wanted to impress the girl by adding an 18% tip to the total cost of the meal. Two entrees came to a total of $21.50. Desserts cost $4.25 and beverages cost $5.75. What is the 18% tip going to cost?

A. $0.78

C. $3.78

B. $1.78

D. $5.67

23) A circular logo is enlarged to fit the lid of a jar. The new diameter is 60% larger than the original. By what percentage has the circumference of the logo increased?

A. 40%

C. 60%

B. 30%

D. 20%

24) All of the following are examples of composite numbers except which one?

A. 2

C. 9

B. 8

D. 14

25) On average, the center for a basketball team hits 25% of his shots while a guard on the same team hits 34% of his shots. If the guard takes 140 shots this year, about how many shots will the center have to take to score as many points as the guard assuming each shot is worth the same number of points?

A. 332

C. 380

B. 190

D. 76

Test 8

Mathematics Knowledge

- o **25 questions**

- o **Total time for this section:** 22 Minutes

- o **Calculators are not allowed at the test.**

1) Which of the following angles is obtuse?

   A. 66 Degrees   C. 45 Degrees

   B. 89 Degrees   D. 134 Degrees

2) What is the product of the square root of 49 and the square root of 36?

   A. 1764   C. 22

   B. 16   D. 42

3) What is the value of 4!?

   A. 120   C. 24

   B. 15   D. 10

4) If a circle has a radius of 29 feet, what's the closest approximation for its circumference?

   A. 87   C. 182

   B. 209   D. 58

5) (6) a^2 = 150, then a is ___ .

   A. a prime number

   B. a negative number

   C. a positive number

   D. either a positive or negative number

6) In the diagram provided, what is the value of a?

A. 220°

B. 140°

C. 40°

D. 320°

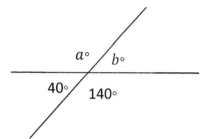

7) Given the diagram, what is the perimeter of the quadrilateral?

A. 620

B. 66

C. 54

D. 33,480

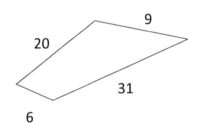

8) $(3x + 3)(x + 5)$

A. $4x + 8$

B. $3x + 3x + 15$

C. $3x^2 + 18x + 15$

D. $3x^2 + 3$

9) Which of the following answer choices lists numbers in increasing order?

A. $-6, 0, 6^1, 6^0$

B. $0, -6, 6^1, 6^0$

C. $6^1, 6^0, 0, -6$

D. $-6, 0, 6^0, 6^1$

10) 75% converted to a fraction equals to:

A. $\dfrac{3}{4}$

B. $\dfrac{15}{25}$

C. $\dfrac{2}{3}$

D. $\dfrac{1}{4}$

11) What is 23,300 in scientific notation?

    A. 7^{56}

    B. 2.33×10^4

    C. 2.33

    D. 23.3

12) The cube root of 64 is?

    A. $3\sqrt{4}$

    B. 2

    C. 4

    D. 444

13) $(4^6)^8 = ?$

    A. -4^2

    B. 4^2

    C. 4^{48}

    D. 256^2

14) The square root of 81?

    A. 81

    B. 8

    C. $2\sqrt{9}$

    D. 9

15) In the circle seen below, there are two chords shown. What is the value of x?

    A. 5

    B. 10

    C. 12

    D. 14

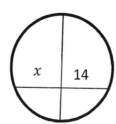

16) What's the greatest common factor of the 25 and 15?

    A. 44

    B. 15

    C. 5

    D. 3

17) What is the value of x?

A. 45°

B. 55°

C. 225°

D. 40°

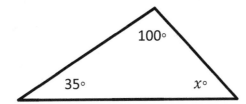

18) What is the expanded equation from of $(5x - 3)(x + 3) = 0$?

A. $5x^2 + 6x + 9$

C. $x^2 + 6x - 9$

B. $5x^2 + 12x - 9$

D. $x^2 + 6x + 9$

19) $0.06 = ?$

A. 0.6%

C. 60.0%

B. 6.0%

D. 600.0%

20) Which of the following is a prime number?

A. 8

C. 12

B. 9

D. 17

21) 58% converted to a fraction.

A. $\frac{2}{5}$

C. $\frac{29}{50}$

B. $\frac{9}{50}$

D. $\frac{2}{50}$

22) How many inches are there in 9 feet?

    A. 108 Inches   C. 54 Inches

    B. 18 Inches   D. 12 Inches

23) $0.68 + 1.26 = ?$

    A. 0.194   C. 19.4

    B. 194   D. 1.94

24) $12\sqrt{2} - 5\sqrt{2}$

    A. 7   C. 17

    B. $7\sqrt{2}$ D. $17\sqrt{2}$

25) What is 9911.16049 rounded to the nearest tenth?

    A. 9911.160

    B. 9911.2

    C. 9911

    D. 9911.16

Test 9

Arithmetic Reasoning

- ○ **25 questions**

- ○ **Total time for this section:** 29 Minutes

- ○ **Calculators are not allowed at the test.**

1) If $x + 5x + 3x = -27$, then what is the value of x?

    A. −3   C. −1

    B. −2   D. 0

2) At Michael's smoothie shop, smoothies cost $2.2 during the week and $3.00 on weekends. If Jonathan bought a smoothie every day except Saturday how much did he spend on smoothies that week?

    A. $14.2   C. $14

    B. $18   D. $10

3) In a bundle of 40 fruits, 8 are apples and the rest are bananas. What percent of the bundle is composed of apples?

    A. 47%   C. 20%

    B. 47.75%   D. 40.75%

4) Which of the following is NOT a factor of 60?

    A. 8   C. 5

    B. 4   D. 15

5) A classroom is a rectangular block that is 50 feet by 10 feet in length and width respectively. If a student walks around the block once, how many yards does the student cover?

    A. 40 Yards   C. 45 Yards

    B. 35 Yards   D. 30 Yards

6) Convert 0.089 to a percent.

    A. 0.9%   C. 8.9%

    B. 0.89%   D. 89%

7) In a bundle of 50 pencils, 12 are red and the rest are blue. What percent of the bundle is composed of blue pencils?

    A. 67%   C. 88%

    B. 84%   D. 76%

8) Emily reported to work at 9 : 15 AM and went out at 7 : 30 PM. How many hours and minutes did she stay in the office?

    A. 10 hours

    B. 7 hours

    C. 10 hours and 15 minutes

    D. 7 hours and 15 minutes

9) A newlywed couple bought a car with $4300 down payment and needs to pay $430 per month for five and half years. How much did they have to pay for that car, all in all?

    A. $32,680   C. $36,600

    B. $33,608   D. $29,180

10) Emily just left for vacation 3 hours ago. She has driven about 70 miles per hour on the 2-lane road and take an exit to get dinner. Then, she called her dad to check in. Her Dad has decided to join her after initially turning down her invitation. How long will Emily have to wait for him if he drives 50 miles per hour?

A. 2 Hours + 10 Minutes

B. 4 Hours + 12 Minutes

C. 4 Hours + 48 Minutes

D. 4 Hours + 30 Minutes

11) There are 72 language students in a language learning center. Currently, 44 students taking Spanish and 22 students taking German. Of the students studying German or Spanish, 8 are taking both courses. How many students are not enrolled in either course?

A. 7   C. 9

B. 24   D. 14

12) A man goes to the grocery store. He buys a carton of milk for $8.20, a dozen eggs for $1.80, and a pound of ground beef for $5. If the tax is 10%, what will his total bill be?

A. $12.90   C. $15.50

B. $15   D. $16.50

13) What is 34,000,000,000 in scientific notation?

A. 3.4×10^{10} C. 0.34×10^{11}

B. 3.4×10^{-10} D. 34×10^{9}

14) Aiden is going to waterproof his carpeting which is 15 feet x 20 feet. The rate is 57 cents per square foot to apply the waterproofing solution. Aiden's total cost will be ___ .

A. $171   C. $173.3

B. $169   D. $100

15) What is the product of the square root of 144 and the square root of 100?

    A. 14,400   C. 100

    B. 60   D. 120

16) In a bundle of 90 pencils, 43 are red and the rest are blue. About what percent of the bundle is composed of blue pencils?

    A. 53%   C. 54%

    B. 50%   D. 52%

17) Daniel is 78 years old, twice as old as Henry. How old is Henry?

    A. 26 years old   C. 29 years old

    B. 34 years old   D. 39 years old

18) Emily lives $5\frac{1}{4}$ miles from where she works. When traveling to work, she walks to a bus stop $\frac{1}{3}$ of the way to catch a bus. How many miles away from her house is the bus stop?

    A. $4\frac{1}{3}$ Miles   C. $2\frac{3}{4}$ Miles

    B. $4\frac{3}{4}$ Miles   D. $1\frac{3}{4}$ Miles

19) If angles A and B are angles of a parallelogram, what is the sum of the measures of the two angles?

    A. 360 degrees   C. 90 degrees

    B. 180 degrees   D. Cannot be determined

20) How much greater is the value of $3x + 9$ than the value of $3x - 5$?

    A. 10   C. 14

    B. 12   D. 16

21) If doughnuts are usually 75 cents each, but there is a sale on Friday advertising them as 1 dozen + ½ dozen for $10.80, what is the new cost for just one?

    A. 50 Cents   C. 60 Cents

    B. 65 Cents   D. 55 Cents

22) What is fourteen percent as a fraction?

    A. $\dfrac{9}{52}$ C. $\dfrac{1}{15}$

    B. $\dfrac{1}{16}$ D. $\dfrac{7}{50}$

23) Ava just bought 4 take–out Mexican dinners. They were priced $13.50, $13.50, $17.25, and $15.35. Ava had a coupon to "buy one, get a second one for half–price on all items that are the same price." (Only 1 coupon per purchase allowed). What is the total amount she paid?

    A. $54.80   C. $54.20

    B. $52.85   D. $60.40

24) School enrollments have fluctuated quite a bit in Oregon this year. The state department of education predicted there would be 200,000 students enrolled state–wide. The enrollment ended up being 250,000. How much of a percentage increase over the projected amount is this?

   A. a 15% increase in the projected amount.

   B. a 25% increase in the projected amount.

   C. a 10% increase in the projected amount.

   D. a 20% increase in the projected amount.

25) There are 53 convention participants staying in the university dormitory. If one dormitory room accommodates 8 participants, how many participants will be staying in the room that is not full?

   A. 7

   B. 6

   C. 5

   D. 4

Test 9

Mathematics Knowledge

- o **25 questions**

- o **Total time for this section:** 22 Minutes

- o **Calculators are not allowed at the test.**

1) What is 2.5×3^3?

    A. 67.5   C. 75

    B. 70.5   D. 675

2) 8 feet, 10 inches + 5 feet, 12 inches = how many inches?

    A. 178 Inches   C. 182 Inches

    B. 188 Inches   D. 200 Inches

3) If $a = 4$ then $a^a \cdot a =$

    A. 40   C. 1,024

    B. 256   D. $4x$

4) Which of the following angles is obtuse?

    A. 20 Degrees   C. 189 Degrees

    B. 40 Degrees   D. 110 Degrees

5) Use the diagram provided as a reference. If the length between point A and C is 87 cm, and the length between point A and B is 46 cm, what is the length between point B and C?

    A. 41 cm

    B. 93 cm

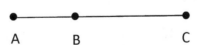

                                A   B   C

    C. 31 cm

    D. 46 cm

6) Liam's average (arithmetic mean) on two mathematics tests is 8. What should Liam's score be on the next test to have an overall of 9 for all the tests?

A. 8

C. 10

B. 9

D. 11

7) What's the area of a trapezoid if the top width is 22 cm, the bottom width is 28 cm, and the height is 10 cm?

A. 250 CM

C. 1000 CM

B. 280 CM

D. 100 CM

8) Which of the following fractions is the largest?

A. $\frac{5}{8}$

C. $\frac{8}{9}$

B. $\frac{3}{7}$

D. $\frac{5}{11}$

9) If $7 + 2x \leq 15$, what is the value of $x \leq$?

A. $14x$

C. -4

B. 4

D. $15x$

10) $(x - 4)(x^2 + 5x + 4) = $?

A. $x^3 + x^2 - 16x + 16$

C. $x^3 + x^2 - 16x - 16$

B. $x^3 + 2x^2 - 16x - 16$

D. $x^3 + x^2 + 16x - 15$

11) Which of the following is not a composite number?

A. 13

C. 24

B. 26

D. 8

12) Simplify $\sqrt{245}$.

    A. $6\sqrt{4}$ C. $7\sqrt{245}$

    B. $7\sqrt{5}$ D. $8\sqrt{245}$

13) If two angles in a triangle measure 53 degrees and 45 degrees, what is the value of the third angle?

    A. 8 Degrees   C. 82 Degrees

    B. 42 Degrees   D. 98 Degrees

14) Given a circle with a radius of 0.55 meters, find the area of the circle.

    A. $1.4\ m^2$ C. $1.9\ m^2$

    B. $1.73\ m^2$ D. $0.95\ m^2$

15) In the figure below, line A is parallel to line B. What is the value of angle x?

    A. 35 Degree

    B. 45 Degree

    C. 100 Degree

    D. 145 Degree

16) How many 3 × 3 squares can fit inside a rectangle with a height of 54 and width of 12?

    A. 72   C. 62

    B. 52   D. 42

17) The cube root of 1,331 is?

    A. $10\sqrt{2}$ C. 11

    B. 10   D. 665.5

18) What is the value of $(12 - 8)!$?

    A. 20   C. 4

    B. 24   D. 28

19) Which of the following pairs are complementary angles?

    A. 50, 40   C. 60, −60

    B. 90, 90   D. 80, 40

20) What is 21,8210 in scientific notation?

    A. 218.21×10^3 C. 0.21821×10^6

    B. 21.821×10^4 D. 2.1821×10^5

21) 12 yard + 10 feet =

    A. 22 Feet   C. 46 Feet

    B. 34 Feet   D. 44 Feet

22) Which of the following is not synonym for 10^2? c

  A. 10 cubed

  B. 10 squared

  C. the square of 10

  D. 10 to the second power

23) What's the greatest common factor of 32 and 48?

    A. 2   C. 8

    B. 12   D. 16

24) If $4^{37} = 4^x \times 4^{20}$, what is the value of x?

    A. 10   C. 20

    B. 1.85   D. 17

25) Which of the following is not a type of quadrilateral?

    A. Parallelogram   C. Trapezoid

    B. Pentagon   D. Square

Test 10

Arithmetic Reasoning

- ○ **25 questions**
- ○ **Total time for this section:** 29 Minutes
- ○ **Calculators are not allowed at the test.**

1) In a classroom of 68 students, 32 are male. About what percentage of the class is female?

   A. 57%   C. 53%

   B. 45%   D. 47%

2) How many square feet of tile is needed for a 20 ft x 20 ft room?

   A. 300 Square Feet   C. 400 Square Feet

   B. 200 Square Feet   D. 160 Square Feet

3) With what number must 9.529769 be multiplied in order to obtain the number 952.9769?

   A. 100 Square Feet   C. 10,000 Square Feet

   B. 1,000 Square Feet   D. 100,000 Square Feet

4) Which of the following is NOT a factor of 70?

   A. 5   C. 6

   B. 7   D. 14

5) Convert 0.0027 to a percent

   A. 0.03%   C. 0.27%

   B. 0.27%   D. 0.027%

6) On average, the center for a basketball team hits 26% of his shots while a guard on the same team hits 50% of his shots. If the guard takes 130 shots this year, how many shots will the center have to take to score as many points as the guard assuming each shot is worth the same number of points?

   A. 65   C. 250

   B. 195   D. 260

7) A machine in a factory has an error rate of 12 parts per 100. The machine normally runs 24 hours a day and produces 80 parts per hour. Yesterday, the machine was shut down for 6 hours for maintenance.

   How many error–free parts did it produce yesterday?

   A. 1,267.2   C. 1,689.6

   B. 230.4   D. 52.4

8) Ava uses a 35% off coupon when buying a sweater that costs $36.12. If she also pays 5% sales tax on the purchase, how much does she pay?

   A. 26.95   C. 21.7

   B. 24.65   D. 34.83

9) A tiger in a zoo has consumed 63 pounds of food in 7 days. If the tiger continues to eat at the same rate, in how many more days will its total food consumption be 99 pounds?

   A. 6   C. 4

   B. 1   D. 2

10) 6 liters of water are poured into an aquarium that's 15cm long, 5cm wide, and 60cm high. How many cm will the water level in the aquarium rise due to this added water? (1 liter of water = 1000 cm^3)

   A. 80   C. 20

   B. 40   D. 10

11) If you invest $1,000 at an annual rate of 9%, how much interest will you earn after one year?

    A. 9   C. 900

    B. 9000   D. 90

12) If a gas tank can hold 45 gallons, how many gallons does it contain when it is $\frac{2}{3}$ full?

    A. 15   C. 67.5

    B. 30   D. 90

13) David bought a 14 pack of soda for $16.60. About how much did he pay per can of soda?

    A. $1.19   C. $4.15

    B. $232.4   D. $14

14) 6 tons of gravel weighs how many pounds?

    A. 3,000   C. 12,000

    B. 6,000   D. 18,000

15) If 9 people can stain a deck in 3 days, how many people would be required to stain the deck in only 1 day?

    A. 3   C. 18

    B. 9   D. 27

16) Ella bought a pair of gloves for $12.49. She gave the clerk $18.00. How much change should she get back?

    A. $4.51   C. $6.51

    B. $5.51   D. $7.51

17) How long will it take to reach a town 360 miles a way if you drive 60 miles per hour?

    A. 6   C. 10

    B. 8   D. 12

18) If a gas tank can hold 25 gallons, how many gallons does it contain when it is $\frac{2}{5}$ full?

    A. 50   C. 62.5

    B. 125   D. 10

19) After being dropped a ball always bounces back to 1/2 of the height of its previous bounce. After the first bounce it reaches a height of 112 inches. How high will it reach on bounce number 3?

    A. 56   C. 14

    B. 28   D. 224

20) If there were a total of 200 raffle tickets sold and you bought 15 tickets, what's the probability that you'll win the raffle?

    A. 7.5%   C. 9.5%

    B. 8.5%   D. 11.5%

21) If the ratio of home fans to visiting fans in a crowd is 5:3 and all 50,000 seats in a stadium are filled, how many home fans are in attendance?

    A. 18,750   C. 10,000

    B. 31,250   D. 28,000

22) What's the next number in the series {5, 11, 17, 23, ?}

    A. 26   C. 28

    B. 27   D. 29

23) What is the result when 12 is added to the product of 4 and 5?

    A. 12   C. 32

    B. 20   D. 24

24) Which of the following is equivalent to 8 times the total of a plus b?

    A. $8a + b$ C. $8(a + b)$

    B. $8b + a$ D. $8(a - b)$

25) What is the absolute value of the quantity six minus nine?

    A. $- 3$ C. $- 15$

    B. 15   D. 3

Test 10

Mathematics Knowledge

- ○ **25 questions**
- ○ **Total time for this section:** 22 Minutes
- ○ **Calculators are not allowed at the test.**

1) The cube of 6 is ___ .

   A. 216

   B. 36

   C. 60

   D. 18

2) If a rectangular swimming pool has a perimeter of 112 feet and is 22 feet wide, what is its area?

   A. 1,496 ft^2

   B. 90 ft^2

   C. 2,464 ft^2

   D. 748 ft^2

3) In the diagram, the straight line is divided by one angled line at 115°. Solve for the value of a.

   A. 65°

   B. 95°

   C. 75

   D. 180°

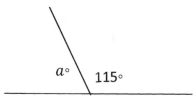

4) $2 - 16 \div (4^2 \div 2) =$ ___

   A. 6

   B. $\frac{3}{4}$

   C. 0

   D. −2

5) $4x^2y^3 + 5x^3y^5 - (5x^2y^3 - 2x^3y^5) =$ ___

   A. $-x^2y^3$

   B. $6x^2y^3 - x^3y^5$

   C. $7x^2Y^3$

   D. $7x^3Y^5 - x^2Y^3$

6) Jacob is having a birthday party for his daughter and is serving orange juice to the 8 children in attendance. If Jacob has 4 liter of orange juice and wants to divide it equally among the children, how many liters does each child get?

A. $\frac{1}{8}$ of litter

C. $\frac{1}{2}$ of litter

B. $\frac{1}{7}$ of litter

D. $\frac{1}{16}$ of litter

7) If x = 6 and y = −2, what is the value of the expression?

$$-x^3 - x^2y + 2y^2 - xy + 4$$

A. −60

C. 168

B. −36

D. −120

8) What is $\sqrt{16} \times \sqrt{81}$

A. 76

C. 36

B. $\sqrt{65}$

D. $\sqrt{13}$

9) Factor $x^2 - 25$

A. $(x - 5)^2$

C. $(x + 5)(x - 5)$

B. $(x + 5)^2$

D. $(x - 5)(x - 5)$

10) $8^6 \times 8^{12}$?

A. 8^{18}

C. 18^8

B. 8^{-6}

D. 8^6

11) What's the circumference of a circle that has a diameter of 15m?

A. 47.124 m

C. 94.25 m

B. 706.9 m

D. 36 m

12) What angle is complementary to 34 degrees?

    A. 46 Degrees   C. 96 Degrees

    B. 56 Degrees   D. 146 Degrees

13) What is the perimeter of the rhombus shown in the figure?

    A. 40 cm

    B. 70 cm

    C. 80 cm

    D. 100 cm

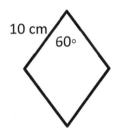

14) What is the mode in this series of numbers: 3, 3, 5, 5, 5, 5, 9, 9, 9, 12, 12, 14

    A. 9   C. 5

    B. 14   D. 12

15) $3.7 \times 10^3 =$ ___

    A. 370   C. 3,700

    B. 37,000   D. 37

16) Given the diagram of parallel lines, what is the value of a?

    A. 301∘

    B. 59∘

    C. 46∘

    D. 247∘

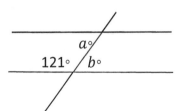

17) Which of the following is an obtuse angle?

    A. 55∘   C. 215∘

    B. 102∘   D. 95∘

18) What's the greatest common factor of 84, 98 and 106?

    A. 18   C. 6

    B. 10   D. 2

19) What is 322,000 in normalized scientific notation?

    A. 3.22×10^5

    B. 322×10^3

    C. 0.322×10^6

    D. 3.22

20) What is the value of 5!

    A. 120   C. 24

    B. 240   D. 720

21) How many inches are there in 10 feet?

    A. 120   C. 80

    B. 100   D. 30

22) 3.93 + 1.74 = ?

    A. 0.567   C. 56.7

    B. 567   D. 5.67

23) What is the mean of the following numbers: 55, 12, 13, 16

    A. 24

    B. 12

    C. 96

    D. 48

24) 1 hour 20 minutes + 3 hours 40 minutes = ?

    A. 2 Hours 10 Minutes

    B. 4 Hours 50 Minutes

    C. 4 Hours 10 Minutes

    D. 5 Hours

25) Which of the following is equivalent to $x^2 + 8x - 9$?

    A. $x\,(x + 9)$

    B. $(x + 1)(x - 9)$

    C. $x\,(x - 9)$

    D. $(x - 1)(x + 9)$

FAOQT Mathematics

Practice Tests

Answers and Explanations

10 Complete Arithmetic and Mathematics Knowledge Tests

Test 1 - AFOQT							
Arithmetic Reasoning				**Mathematics Knowledge**			
1)	B	16)	C	1)	C	16)	B
2)	D	17)	C	2)	D	17)	B
3)	D	18)	D	3)	B	18)	C
4)	B	19)	B	4)	A	19)	C
5)	A	20)	D	5)	C	20)	A
6)	C	21)	D	6)	D	21)	A
7)	A	22)	A	7)	B	22)	A
8)	A	23)	A	8)	B	23)	B
9)	C	24)	D	9)	D	24)	C
10)	D	25)	D	10)	A	25)	A
11)	C			11)	C		
12)	A			12)	B		
13)	B			13)	C		
14)	C			14)	C		
15)	B			15)	B		

Test 2 - AFOQT							
Arithmetic Reasoning				**Mathematics Knowledge**			
1)	B	16)	C	1)	C	16)	A
2)	B	17)	C	2)	D	17)	B
3)	C	18)	C	3)	D	18)	D
4)	C	19)	C	4)	D	19)	B
5)	B	20)	C	5)	B	20)	A
6)	D	21)	D	6)	C	21)	D
7)	B	22)	B	7)	B	22)	A
8)	A	23)	B	8)	C	23)	B
9)	C	24)	A	9)	C	24)	D
10)	C	25)	D	10)	B	25)	B
11)	A			11)	C		
12)	B			12)	D		
13)	C			13)	D		
14)	B			14)	A		
15)	B			15)	C		

Test 3 - AFOQT							
Arithmetic Reasoning				**Mathematics Knowledge**			
1)	C	16)	C	1)	C	16)	C
2)	D	17)	D	2)	C	17)	A
3)	C	18)	D	3)	A	18)	D
4)	A	19)	A	4)	D	19)	C
5)	C	20)	C	5)	B	20)	C
6)	C	21)	C	6)	A	21)	A
7)	A	22)	B	7)	C	22)	B
8)	D	23)	C	8)	D	23)	D
9)	C	24)	A	9)	B	24)	C
10)	D	25)	B	10)	B	25)	C
11)	A			11)	A		
12)	C			12)	A		
13)	D			13)	C		
14)	D			14)	B		
15)	C			15)	A		

Test 4 - AFOQT							
Arithmetic Reasoning				Mathematics Knowledge			
1)	C	16)	C	1)	B	16)	A
2)	D	17)	B	2)	A	17)	C
3)	C	18)	D	3)	C	18)	C
4)	D	19)	B	4)	B	19)	C
5)	B	20)	B	5)	D	20)	C
6)	B	21)	A	6)	B	21)	B
7)	D	22)	A	7)	A	22)	D
8)	C	23)	C	8)	A	23)	C
9)	A	24)	B	9)	D	24)	C
10)	D	25)	C	10)	B	25)	D
11)	D			11)	A		
12)	B			12)	C		
13)	C			13)	D		
14)	D			14)	D		
15)	A			15)	A		

Test 5 - AFOQT							
Arithmetic Reasoning				**Mathematics Knowledge**			
1)	D	16)	D	1)	D	16)	B
2)	B	17)	A	2)	A	17)	D
3)	A	18)	A	3)	A	18)	C
4)	C	19)	B	4)	D	19)	A
5)	B	20)	D	5)	A	20)	B
6)	A	21)	B	6)	D	21)	A
7)	A	22)	A	7)	C	22)	D
8)	D	23)	B	8)	B	23)	B
9)	A	24)	D	9)	D	24)	C
10)	C	25)	C	10)	A	25)	C
11)	C			11)	B		
12)	B			12)	C		
13)	B			13)	D		
14)	C			14)	A		
15)	C			15)	B		

Test 6 - AFOQT

Arithmetic Reasoning				Mathematics Knowledge			
1)	D	16)	C	1)	D	16)	C
2)	D	17)	A	2)	D	17)	C
3)	C	18)	C	3)	C	18)	D
4)	C	19)	A	4)	A	19)	B
5)	A	20)	B	5)	A	20)	B
6)	A	21)	C	6)	C	21)	C
7)	B	22)	D	7)	C	22)	C
8)	D	23)	A	8)	B	23)	A
9)	A	24)	D	9)	A	24)	D
10)	A	25)	C	10)	A	25)	C
11)	C			11)	A		
12)	A			12)	C		
13)	B			13)	D		
14)	B			14)	C		
15)	B			15)	B		

Test 7 - AFOQT							
Arithmetic Reasoning				**Mathematics Knowledge**			
1)	B	16)	B	1)	C	16)	B
2)	C	17)	D	2)	D	17)	A
3)	C	18)	B	3)	C	18)	B
4)	D	19)	A	4)	D	19)	C
5)	A	20)	A	5)	D	20)	A
6)	B	21)	B	6)	C	21)	C
7)	B	22)	D	7)	C	22)	B
8)	D	23)	C	8)	D	23)	A
9)	D	24)	A	9)	B	24)	B
10)	B	25)	B	10)	B	25)	D
11)	A			11)	C		
12)	A			12)	A		
13)	D			13)	A		
14)	D			14)	D		
15)	B			15)	B		

Test 8 - AFOQT							
Arithmetic Reasoning				**Mathematics Knowledge**			
1)	A	16)	B	1)	D	16)	C
2)	D	17)	C	2)	D	17)	A
3)	A	18)	B	3)	C	18)	B
4)	B	19)	D	4)	C	19)	B
5)	C	20)	D	5)	D	20)	D
6)	D	21)	C	6)	B	21)	C
7)	B	22)	D	7)	B	22)	A
8)	D	23)	B	8)	C	23)	D
9)	C	24)	A	9)	D	24)	B
10)	A	25)	B	10)	A	25)	B
11)	C			11)	B		
12)	D			12)	C		
13)	B			13)	C		
14)	A			14)	C		
15)	D			15)	D		

Test 9 - AFOQT						
Arithmetic Reasoning				**Mathematics Knowledge**		
1)	A	16)	D	1)	A	16)	A
2)	C	17)	D	2)	A	17)	C
3)	C	18)	D	3)	C	18)	B
4)	A	19)	D	4)	D	19)	A
5)	A	20)	C	5)	A	20)	D
6)	C	21)	C	6)	D	21)	C
7)	D	22)	D	7)	A	22)	A
8)	C	23)	B	8)	C	23)	D
9)	A	24)	B	9)	B	24)	D
10)	B	25)	C	10)	C	25)	B
11)	D			11)	A		
12)	D			12)	B		
13)	A			13)	C		
14)	A			14)	D		
15)	D			15)	D		

Test 10 - AFOQT							
Arithmetic Reasoning				**Mathematics Knowledge**			
1)	C	16)	B	1)	A	16)	B
2)	C	17)	A	2)	D	17)	B
3)	A	18)	D	3)	A	18)	D
4)	C	19)	B	4)	C	19)	A
5)	B	20)	A	5)	D	20)	A
6)	C	21)	B	6)	C	21)	A
7)	A	22)	D	7)	D	22)	D
8)	B	23)	C	8)	C	23)	A
9)	C	24)	C	9)	C	24)	D
10)	A	25)	D	10)	A	25)	D
11)	D			11)	A		
12)	B			12)	B		
13)	A			13)	A		
14)	C			14)	C		
15)	D			15)	C		

AFOQT Math Practice Tests Explanations

In this section, answers and explanations are provided for two AFOQT Practice Math Tests, Test 1 and Test 6. Review the answers and explanations to learn more about solving math questions fast.

Test 1 Arithmetic Reasoning

Answers and Explanations

1) **Choice B is correct**

$36 \div 3 = 12$ hours for one course

$12 \times 25 = 300 \Rightarrow \$300$

2) **Choice D is correct**

Michelle = Karen − 9

Michelle = David − 4

Karen + Michelle + David = 82

Karen + 9 = Michelle $\Rightarrow$ Karen = Michelle − 9

Karen + Michelle + David = 82

Now, replace the ages of Karen and David by Michelle. Then:

Michelle + 9 + Michelle + Michelle + 4 = 82

3Michelle + 13 = 82 $\Rightarrow$ 3Michelle = 82 − 13

3Michelle = 69

Michelle = 23

3) **Choice D is correct**

$$distance = speed \times time \Rightarrow \text{time} = \frac{distance}{speed} = \frac{600}{50} = 12$$

(Round trip means that the distance is 600 miles)

The round trip takes 12 hours. Change hours to minutes, then:

$$12 \times 60 = 720$$

4) Choice B is correct

Since Julie gives 8 pieces of candy to each of her friends, then, then number of pieces of candies must be divisible by 8.

a. $187 \div 8 = 23.375$
b. $216 \div 8 = 27$
c. $343 \div 8 = 42.875$
d. $223 \div 8 = 27.875$

Only choice b gives a whole number.

5) Choice A is correct

Area of a rectangle = width × length = 30 × 45 = 1,350

6) Choice C is correct

$$average = \frac{sum}{total}$$

Sum = 2 + 5 + 22 + 28 + 32 + 35 + 35 + 33 = 192

Total number of numbers = 8

$average = \frac{192}{8} = 24$

7) Choice A is correct

The base rate is $15.

The fee for the first 40 visits is: $40 \times 0.20 = 8$

The fee for the visits 41 to 60 is: $20 \times 0.10 = 2$

Total charge: 15 + 8 + 2 = 25

8) Choice A is correct

$$average = \frac{sum}{total} = \frac{32 + 35 + 29}{3} = \frac{96}{3} = 32$$

9) Choice C is correct

The amount they have = $10.25 + $11.25 + $18.45 = 39.95

10) Choice D is correct

Change 9 hours to minutes, then: $9 \times 60 = 540$ minutes

$$\frac{540}{90} = 6$$

11) Choice C is correct

15 dozen of magazines are 180 magazines: $15 \times 12 = 180$

$180 - 57 = 123$

12) Choice A is correct

$$probability = \frac{desired\ outcomes}{possible\ outcomes} = \frac{4}{4 + 3 + 7 + 10} = \frac{4}{24} = \frac{1}{6}$$

13) Choice B is correct

Find the value of each choice:

A. $2 \times 2 \times 5 \times 7 = 140$

B. $2 \times 2 \times 2 \times 2 \times 5 \times 7 = 560$

C. $2 \times 7 = 14$

D. $2 \times 2 \times 2 \times 5 \times 7 = 280$

14) Choice C is correct

1 ton = 2,000 pounds

5 ton = 10,000 pounds

$$\frac{32,000}{10,000} = 3.2$$

William needs to make at least 4 trips to deliver all of the food.

15) Choice B is correct

$180 - 40 - 50 = 90$

16) Choice C is correct

Each worker can walk 3 dogs: $9 \div 3 = 3$

5 workers can walk 15 dogs.

$5 \times 3 = 15$

17) Choice C is correct

2 weeks = 14 days

$14 \times 3 = 42$ hours

$42 \times 60 = 2{,}520$ minutes

18) Choice D is correct

$\dfrac{180}{20} = 9$

19) Choice B is correct

$30\% \times 50 = \dfrac{30}{100} \times 50 = 15$

The coupon has $15 value. Then, the selling price of the sweater is $35.

$50 - 15 = 35$, Add 5% tax, then: $\dfrac{5}{100} \times 35 = 1.75$ for tax, Total: $35 + 1.75 = \$36.75$

20) Choice D is correct

1 quart = 0.25 gallon

34 quarts = $34 \times 0.25 = 8.5$ gallons

then: $\dfrac{8.5}{2} = 4.25$ weeks

21) Choice D is correct

The difference of the file added, and the file deleted is:

652,159 – 599,986 = 52,173

937,036 + 52,173 = 989,209

22) Choice A is correct

Write proportion and solve.

$\frac{1}{153} = \frac{32}{x} \Rightarrow x = 32 \times 153 = 4{,}896$

23) Choice A is correct

25 students did not have to go to summer school.

30 – 5 = 25

$\frac{25}{30} = \frac{5}{6}$

24) Choice D is correct

$$average = \frac{sum}{total} = \frac{240}{60} = 4$$

25) Choice D is correct

Write a proportion and solve.

$\frac{1/5}{1} = \frac{50}{x}$ $x = \frac{50}{1/5} = 250$

Test 1 Mathematics Knowledge

Answers and Explanations

1) Choice C is correct

If a = 3 then:

$$b = \frac{a^2}{3} + 3 \Rightarrow b = \frac{3^2}{3} + 3 = 3 + 3 = 6$$

2) Choice D is correct

$$\sqrt[8]{256} = 2$$

$$(2^8 = 2 \times 2 \times 2 \times 2 \times 2 \times 2 \times 2 \times 2 = 256)$$

3) Choice B is correct

(r = radius)

Area of a circle = $\pi r^2 = \pi \times (5)^2 = 3.14 \times 25 = 78.5$

4) Choice A is correct

$$-8a = 64 \quad \Rightarrow \quad a = \frac{64}{-8} = -8$$

5) Choice C is correct

All angles in a triable add up to 180 degrees.

$90° + 45° = 135°$

$x = 180° - 135° = 45°$

6) Choice D is correct

Use Pythagorean Theorem: $a^2 + b^2 = c^2$

$(4)^2 + (5)^2 = c^2 \quad \Rightarrow \quad 16 + 25 = 41 = C^2$

$C = \sqrt{41} = 6.403$

7) Choice B is correct

Use FOIL (first, out, in, last) method.

$(5x + 5)(2x + 6) = 10x^2 + 30x + 10x + 30 = 10x^2 + 40x + 30$

8) Choice B is correct

$$5(a - 6) = 22 \Rightarrow 5a - 30 = 22 \Rightarrow 5a = 22 + 30 = 52$$

$$\Rightarrow 5a = 52 \Rightarrow a = \frac{52}{5} = 10.4$$

9) Choice D is correct

Use exponent multiplication rule:

$$x^a \cdot x^b = x^{a+b}$$

Then:

$$3^{24} = 3^8 \times 3^x = 3^{8+x}$$

$$24 = 8 + x \Rightarrow x = 24 - 8 = 16$$

10) Choice A is correct

An obtuse angle is an angle of greater than 90 degrees and less than 180 degrees. Only choice a is an obtuse angle.

11) Choice C is correct

To factor the expression $x^2 + 5 - 6$, we need to find two numbers whose sum is 5 and their product is -6.

Those numbers are 6 and -1. Then:

$$x^2 + 5 - 6 = (x + 6)(x - 1)$$

12) Choice B is correct

Slope of a line: $\dfrac{y_2 - y_1}{x_2 - x_1} = \dfrac{rise}{run}$

$$\frac{y_2 - y_1}{x_2 - x_1} = \frac{3 - 7}{5 - 6} = \frac{-4}{-1} = 4$$

13) Choice C is correct

$\sqrt{100} = 10$, $\sqrt{36} = 6$

$10 \times 6 = 60$

14) Choice C is correct

Only choice c is not equal to 5^2

15) Choice B is correct

$\sqrt[3]{2,197} = 13$

16) Choice B is correct

In scientific notation form, numbers are written with one whole number times 10 to the power of a whole number. Number 952,710 has 6 digits. Write the number and after the first digit put the decimal point. Then, multiply the number by 10 to the power of 5 (number of remaining digits). Then:

$952,710 = 9.5271 \times 10^5$

17) Choice B is correct

The area of the non-shaded region is equal to the area of the bigger rectangle subtracted by the area of smaller rectangle.

Area of the bigger rectangle = $12 \times 16 = 192$

Area of the smaller rectangle = $10 \times 4 = 40$

Area of the non-shaded region = $192 - 40 = 152$

18) Choice C is correct

$$\sqrt{16x^2} = \sqrt{16} \times \sqrt{x^2} = 4 \times x = 4x$$

19) Choice C is correct

Area of a square = (one side)^2 $\Rightarrow$ A = $(4.5)^2$ $\Rightarrow$ A = 20.25

20) Choice A is correct

Diameter = 2r ⇒ 9 = 2r ⇒ r = 4.5

Circumference = 2πr ⇒ C = 2π(4.5) ⇒ C = 9π

21) Choice A is correct

Diameter = 2r ⇒ 2.5 = 2r ⇒ r = 1.25

Circumference = 2πr ⇒ C = 2π(1.25) ⇒ C = 2.5π

22) Choice A is correct

Factor of 76: {1, 2, 4, 19, 38, 76}

Factor of 20: {1, 2, 4, 5, 10, 20}

Then, factors they have in common is {1, 2, 4}

23) Choice B is correct

Diameter = 2r ⇒ 22 = 2r ⇒ r = 11

Circumference = 2πr ⇒ C = 2π(11) ⇒ C = 22 × 3.14 = 69.08

24) Choice C is correct

The straight line is 180 degrees. Then:

a = 180∘ − 115∘ = 65∘

25) Choice A is correct

Volume = length × width × height

Volume = 9 × 6 × 9

Volume = 486 cm³

Test 6 Arithmetic Reasoning

Answers and Explanations

1) Choice D is correct

2 weeks = 14 days

Then: 14 × 6 = 84 hours

84 × 60 = 5,040 minutes

2) Choice D is correct

$$distance = speed \times time \Rightarrow \text{time} = \frac{distance}{speed} = \frac{340 + 340}{50} = 13.6$$

(Round trip means that the distance is 680 miles)

The round trip takes 13.6 hours. Change hours to minutes, then: $13.6 \times 60 = 816$

3) Choice C is correct

60 − 42 = 18 male students

$\frac{18}{60} = 0.3$

Change 0.3 to percent ⇒ 0.3 × 100 = 30%

4) Choice C is correct

$$average = \frac{sum}{total}$$

Sum = 7 + 9 + 22 + 28 + 28 + 30 = 124

Total number of numbers = 9

$\frac{124}{6} = 20.67$

5) Choice A is correct

Emma's three best times are 54, 57, and 57.

The average of these numbers is:

$$average = \frac{sum}{total}$$

Sum = 54 + 57 + 57 = 168

Total number of numbers = 3

$$average = \frac{168}{3} = 56$$

6) Choice A is correct

The area of a 15 feet x 15 feet room is 225 square feet.
15 × 15 = 225

7) Choice B is correct

1.303572 × 1000 = 1303.572

8) Choice D is correct

The factors of 50 are:

{1, 2, 5, 10, 25, 50}

15 is not a factor of 50.

9) Choice A is correct

4 percent of 25 is: $25 \times \frac{4}{100} = 1$

Emma's new rate is 26.

25 + 1= 26

10) Choice A is correct

Emily = Lucas

Emily = 4 Mia ⇒ Lucas = 4 Mia

Lucas = Mia + 21

then: Lucas = Mia + 21 ⇒ 4 Mia = Mia + 21

Remove 1 Mia from both sides of the equation. Then:

3 Mia = 21 ⇒ Mia = 7

11) Choice C is correct

12 days, 12 × 5 = 60 hours, 60 × 60 = 3,600 minutes

12) Choice A is correct

Sum = 22 + 34 + 16 + 20 = 92

$$average = \frac{92}{4} = 23$$

13) Choice B is correct

Perimeter of a rectangle = 2 × length + 2 × width =

2 × 90 + 2 × 30 = 180 + 60 = 240

14) Choice B is correct

$$Speed = \frac{\text{distance}}{\text{time}}$$

$$16.2 = \frac{distance}{2.1} \Rightarrow distance = 16.2 \times 2.1 = 34.02$$

Rounded to a whole number, the answer is 34.

15) Choice B is correct

Let's review the choices provided and find their sum.

a. 20 × 6 = 120
b. 26 × 6 = 144 ⇒ is greater than 120 and less than 180
c. 30 × 6 = 180
d. 34 × 6 = 204

Only choice b gives a number that is greater than 120 and less than 180.

16) Choice C is correct

$$\frac{1 \ hour}{15 \ coffees} = \frac{x}{1500} \qquad \Rightarrow 15 \times x = 1 \times 1,500 \Rightarrow 15x = 1,500$$

$$x = 100$$

It takes 100 hours until she's made 1,500 coffees.

17) Choice A is correct

$120 - 12 = 108$

$$\frac{108}{12} = 9$$

18) Choice C is correct

$$percent\ of\ change = \frac{change}{original\ number}$$

$7.75 - 7.50 = 0.25$

$$percent\ of\ change = \frac{0.25}{7.50} = 0.0333 \qquad \Rightarrow 0.0333 \times 100 = 3.33\%$$

19) Choice A is correct

Write a proportion and solve.

$$\frac{\frac{1}{2}\,inches}{4.5} = \frac{1\,mile}{x}$$

Use cross multiplication, then:

$$\frac{1}{2}x = 4.5 \rightarrow x = 9$$

20) Choice B is correct

Two candy bars costs 50¢ and a package of peanuts cost 75¢ and a can of cola costs 50¢. The total cost is: 50 + 75 + 50 = 175

175 is equal to 7 quarters. 7 × 25 = 175

21) Choice C is correct

Every day the hour hand of a watch makes 2 complete rotation. Thus, it makes 16 complete rotations in 8 days.

2 × 8 = 16

22) Choice D is correct

$\sqrt{81} \times \sqrt{25} = 9 \times 5 = 45$

23) Choice A is correct

$2y + 4y + 2y = -24 \qquad \Rightarrow 8y = -24 \quad \Rightarrow y = -\dfrac{24}{8} \quad \Rightarrow y = -3$

24) Choice D is correct

$2\dfrac{2}{3} - 1\dfrac{5}{6} = 2\dfrac{4}{6} - 1\dfrac{5}{6} = \dfrac{16}{6} - \dfrac{11}{6} = \dfrac{5}{6}$

25) Choice C is correct

To convert a decimal to percent, multiply it by 100 and then add percent sign (%).
$0.023 \times 100 = 2.30\%$

Test 6 Mathematics Knowledge

Answers and Explanations

1) Choice D is correct

Use FOIL (First, Out, In, Last) method.

$(x + 7)(x + 5) = x^2 + 5x + 7x + 35 = x^2 + 12x + 3$

2) Choice D is correct

In scientific notation form, numbers are written with one whole number times 10 to the power of a whole number. Number 670,000 has 6 digits. Write the number and after the first digit put the decimal point. Then, multiply the number by 10 to the power of 5 (number of remaining digits). Then: $670,000 = 6.7 \times 10^5$

3) Choice C is correct

Perimeter of a triangle = side 1 + side 2 + side 3 = 25 + 25 + 25 = 75

4) Choice A is correct

From the choices provided, 36, 48 and 54 are divisible by 6. From these numbers, 54 is the biggest.

5) Choice A is correct

Oven 1 = 4 oven 2

If Oven 2 burns 3 then oven 1 burns 12 pizzas. 3 + 12 = 15

6) Choice C is correct

An obtuse angle is an angle of greater than 90° and less than 180°.

7) Choice C is correct

Use exponent multiplication rule:

$x^a \cdot x^b = x^{a + b}$

Then: $7^7 \times 7^8 = 7^{15}$

8) Choice B is correct

5231.48245 rounded to the nearest tenth equals 5231.5

(Because 5231.48 is closer to 5,231.5 than 5,231.4)

9) Choice A is correct

$\sqrt[3]{512} = 8$

10) Choice A is correct

Diameter = 16

then: Radius = 8

Area of a circle = πr^2 $\Rightarrow$ A = 3.14(8)^2 = 200.96

11) Choice A is correct

$7! = 7 \times 6 \times 5 \times 4 \times 3 \times 2 \times 1$

12) Choice C is correct

Let's review the choices provided. Put the values of x and y in the equation.

A. (1, 2) $\Rightarrow x = 1 \Rightarrow y = 2$ This is true!

B. (−2, −13) $\Rightarrow x = -2 \Rightarrow y = -13$ This is true!

C. (3, 18) $\Rightarrow x = 3 \Rightarrow y = 12$ This is not true!

D. (2, 7) $\Rightarrow x = 2 \Rightarrow y = 7$ This is true!

13) Choice D is correct

$1 - (-8) = 1 + 8 = 9$

14) Choice C is correct

Use distance formula:

$$d = \sqrt{(x_1 - x_2)^2 + (y_1 - y_2)^2} = \sqrt{(1 - (-2))^2 + (3 - 7)^2}$$

$$\sqrt{9 + 16} = \sqrt{25} = 5$$

15) Choice B is correct

$x^2 - 81 = 0$ $\Rightarrow$ $x^2 = 81$ $\Rightarrow x$ could be 9 or –9.

16) Choice C is correct

Area of a rectangle = width × length = 160 × 200 = 3,200

17) Choice C is correct

Number 2.103119 should be multiplied by 10,000 in order to obtain the number 21,031.19

2.103119 × 10,000 = 21,031.19

18) Choice D is correct

factor of 50 = {1, 2, 5, 10, 25, 50}

100 is not a factor of 50.

19) Choice B is correct

Let's review the choices provided.

A. 80 × 4 = 320

B. 85 × 4 = 340

C. 90 × 4 = 360

D. 95 × 4 = 380

From choices provided, only 340 is greater than 320 and less than 360.

20) Choice B is correct

The cube of 4 = 4 × 4 × 4 = 64

$\frac{1}{4} \times 64 = 16$

21) Choice C is correct

From the list of numbers, 11, 13, and 19 are prime numbers. Their sum is:

11 + 13 + 19 = 43

22) Choice C is correct

$$25\% = \frac{25}{100} = \frac{1}{4}$$

23) Choice A is correct

Two Angles are supplementary when they add up to 180 degrees.

135° + 45° = 180°

24) Choice D is correct

$$\frac{20}{100} \times 50 = 10$$

25) Choice C is correct

$$5(2x^6)^3 \implies 5 \times 2^3 \times x^{18} = 40x^{18}$$

"Effortless Math" Publications

Effortless Math authors' team strives to prepare and publish the best quality Mathematics learning resources to make learning Math easier for all. We hope that our publications help you or your student Math in an effective way.

We all in Effortless Math wish you good luck and successful studies!

Effortless Math Authors

www.EffortlessMath.com

… So Much More Online!

- ✓ FREE Math lessons

- ✓ More Math learning books!

- ✓ Mathematics Worksheets

- ✓ Online Math Tutors

Need a PDF version of this book?

Visit www.EffortlessMath.com

Or send email to: info@EffortlessMath.com

CPSIA information can be obtained
at www.ICGtesting.com
Printed in the USA
BVHW012100210920
589336BV00013B/486